Handbook of PARTICLE SWARM OPTIMIZATION
Concepts, Principles and Applications

Handbook of
PARTICLE SWARM OPTIMIZATION
Concepts, Principles and Applications

Hoveidi J. Bidhendi

Oxford Book Company
Jaipur ▪ New Delhi

Published By

Oxford Book Company

Regd. Off. "Bony Residency", Gate No. 2
Opp. Tilak Public School, Vishweswariya Nagar,
Gopalpura Road, Jaipur - 302018, India
Ph: 0141 - 2761280, Telefax: 2761381
email - oxfordbookjpr@gmail.com
website - oxfordbookcompany.com

Delhi Office: 102, 1st Floor, Satyam House,
4327/3, Ansari Road, Darya Ganj,
New Delhi - 110002, India
Ph: 011-45652440

Distributed by
KEYA INTERNATIONAL PRESS LLP
E-mail: keyainternationalpressllp@gmail.com

First published in India, 2024

ISBN : 9789355240767

Typeset by Shivangi Computers, Jaipur

Printed and bound in India by Replika Press Pvt. Ltd.

Contents

Preface *vii*

1. **Introduction** **1**
Mathematical Optimisation • Classification of Critical Points and Extrema • Metaheuristic • Function of a Real Variable • Particle Swarm Optimisation

2. **Important Terms and Persons** **25**
James Kennedy (Social Psychologist) • Social Behavior • Flocking (Behavior) • Swarm Intelligence • Stochastic Diffusion Search • Gradient • Gradient and the Derivative or Differential • Gradient Descent • Candidate Solution • Meta-optimization • Multi-objective Optimization • Pareto Efficiency • Pareto Improvements and Microeconomic Theory

3. **Type of Swarming** **65**
Swarm Behaviour • Agent-based Model • Ant Colony Optimization Algorithms • Ant Robotics • Artificial Ants • Bees Algorithm • Crowd Simulation • Firefly Algorithm • Glowworm Swarm Optimization • Self-propelled Particles • Swarm (Simulation)

4. **Concepts and Principles** **121**
School of Thought • Mindset • Limit of a Sequence • Modes of Convergence • Premature Convergence • Local Optimum • Convergence (Evolutionary Computing) • Multi-swarm Optimization • Occam's Razor • Particle Filter • Artificial Immune System • Evolutionary Computation • Evolutionary Programming • Evolution Strategy • Eagle Strategy • Cuckoo Search

5. **Computational Complexity Theory** **181**
Computational Problem • Decision Problem • Function Problem • Turing Machine • Additional Details Required to Visualize or Implement Turing Machines • Models Equivalent to the Turing Machine Model • Limitations of Turing Machines • Turing Machine Equivalents

6. Random Access Machine **216**

Introduction to the Random Access Machine Model (RAM) • Bounded Indirection and the Primitive Recursive Functions • Non-deterministic Turing Machine • Best, Worst and Average Case • Quicksort • Implementation Issues • Upper and Lower Bounds on the Complexity of Problems • Big O notation • Complexity Classes • Boolean Circuit

7. Time Hierarchy Theorem **267**

Space Hierarchy Theorem • Reduction (Complexity) • Polynomial-Time Reduction • Log-space Reduction • Complete (Complexity) • NP-complete • P Versus NP Problem • Consequences of the Resolution of the Problem

8. Complexity Classes **309**

Sharp-P • Sharp-P-complete • 2-EXPTIME • AC0 • ACC0 • AC (Complexity) • Arithmetical Hierarchy • Bounded-error Probabilistic Polynomial • Parameterized Complexity

Bibliography 331

Index 333

Preface

Particle Swarm Optimization (PSO) is a recently proposed algorithm by James Kennedy and R. C. Eberhart in 1995, motivated by social behaviour of organisms such as bird flocking and fish schooling. PSO algorithm is not only a tool for optimization , but also a tool for representing sociocognition of human and artificial agents, based on principles of social psychology. Some scientists suggest that knowledge is optimized by social interraction and thinking is not only private but also interpersonal. PSO as an optimization tool, provides a population-based search procedure in which individuals called particles change their position (state) with time. In a PSO system, particles fly around in a multidimensional search space. During flight, each particle adjusts its position according to its own experience, and according to the experience of a neighbouring particle, making use of the best position encountered by itself and its neighbour. Thus, as in modern Gas and memetic algorithms, a PSO system combines local search methods with global search methods, attempting to balance exploration and exploitation.

PSO optimizes a problem by having a population of candidate solutions, here dubbed particles, and moving these particles around in the search-space according to simple mathematical formulae over the particle's position and velocity. Each particle's movement is influenced by its local best known position and is also guided toward the best known positions in the search-space, which are updated as better positions are found by other particles. This is expected to move the swarm toward the best solutions. PSO is originally attributed to Kennedy, Eberhart and Shi and was first intended for simulating social behaviour, as a stylized representation of the movement of organisms in a bird flock or fish school. The algorithm was simplified and it was observed to be performing optimization. The book by Kennedy and Eberhart describes many philosophical aspects of PSO and swarm intelligence. An extensive survey of PSO applications is made by Poli. PSO is a metaheuristic as it makes few or no assumptions about the problem being optimized and can search very large spaces of candidate solutions. However, metaheuristics such as PSO do not guarantee an optimal solution is ever found. More specifically, PSO does not use the gradient of the problem being optimized, which means PSO does not require that the optimization problem bedifferentiable as is required by classic optimization methods such as gradient descent and quasi-newton methods. PSO can therefore also be used on optimization problems that are partially irregular, noisy, change over time, etc. The basic PSO is easily trapped into a local minimum. This premature convergence can be avoided by not using any more the entire swarm's best known position g but just the best known position l of a sub-swarm "around" the particle that is moved. Such a sub-swarm can be a geometrical one - for example "the *m* nearest particles" - or, more often, a social one, i.e. a set of particles that is not depending on any distance. In such a case, the PSO variant is said to be local best.

If we suppose there is an information link between each particle and its neighbours, the set of these links builds a graph, a communication network, that is called the topology of the PSO variant. A commonly used social topology is the ring, in which each particle has just two neighbours, but there are far more. The topology is not necessarily fixed, and can be adaptive here are several schools of thought as to why and how the PSO algorithm can perform optimization. A common belief amongst researchers is that the swarm behaviour varies between exploratory behaviour, that is, searching a broader region of the search-space, and exploitative behaviour, that is, a locally oriented search so as to get closer to a (possibly local) optimum. This school of thought has been prevalent since the inception of PSO. This school of thought contends that the PSO algorithm and its parameters must be chosen so as to properly balance between exploration and exploitation to avoid premature convergence to a local optimum yet still ensure a good rate of convergence to the optimum. Another school of thought is that the behaviour of a PSO swarm is not well understood in terms of how it affects actual optimization performance, especially for higher dimensional search-spaces and optimization problems that may be discontinuous, noisy, and time-varying. This school of thought merely tries to find PSO algorithms and parameters that cause good performance regardless of how the swarm behaviour can be interpreted in relation to e.g. exploration and exploitation.

This Handbook describes many philosophical aspects of PSO and swarm intelligence. PSO is a metaheuristic as it makes few or no assumptions about the problem being optimized and can search very large spaces of candidate solutions.

—*Hoveidi J. Bidhendi*

Chapter 1

Introduction

Mathematical Optimisation

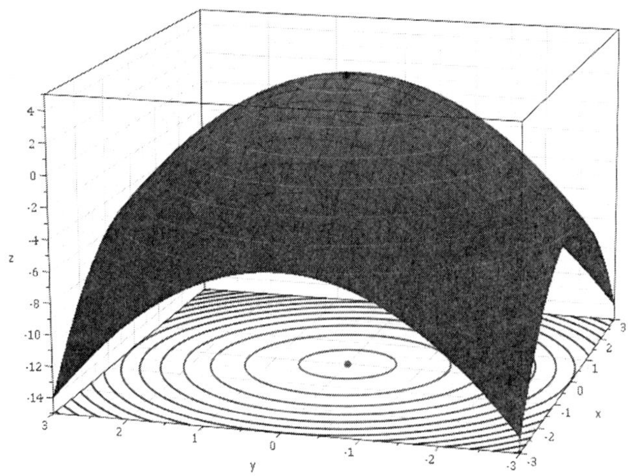

Figure: In mathematics, computational science, or management science, mathematical optimisation (alternatively, optimisation or mathematical programming) refers to the selection of a best element from some set of available alternatives.

In the simplest case, an optimisation problem consists of maximising or minimising a real function by systematically choosing input values from within an allowed set and computing the value of the function.

The generalisation of optimisation theory and techniques to other formulations comprises a large area of applied mathematics. More generally, optimisation includes finding "best available" values of some objective function given a defined domain, including a variety of different types of objective functions and different types of domains.

Optimisation Problems

An optimisation problem can be represented in the following way

Given: a function $f : A \to R$ from some set A to the real numbers

Sought: an element x_0 in A such that $f(x_0)$ d– $f(x)$ for all x in A ("minimisation") or such that $f(x_0)$ e–$f(x)$ for all x in A ("maximisation").

Such a formulation is called an optimisation problem or a mathematical programming problem, a term not directly related to computer programming, but still in use for example in linear programming. Many real-world and theoretical problems may be modelled in this general framework. Problems formulated using this technique in the fields of physics and computer vision may refer to the technique as energy minimisation, speaking of the value of the function f as representing the energy of the system being modelled.

Typically, A is some subset of the Euclidean space R^n, often specified by a set of *constraints*, equalities or inequalities that the members of A have to satisfy. The domain A of f is called the *search space* or the *choice set*, while the elements of A are called *candidate solutions* or *feasible solutions.*

The function f is called, variously, an objective function, cost function (minimisation), utility function (maximisation), or, in certain fields, energy function, or energy functional. A feasible solution that minimises (or maximises, if that is the goal) the objective function is called an *optimal solution.*

By convention, the standard form of an optimisation problem is stated in terms of minimisation. Generally, unless both the objective function and the feasible region are convex in a minimisation problem, there may be several local minima, where a *local minimum* x^* is defined as a point for which there exists some $\delta > 0$ so that for all x such that

$$\|\mathbf{x} - \mathbf{x}^*\| \leq \delta;$$

the expression

$$f(\mathbf{x}^*) \leq f(\mathbf{x})$$

holds; that is to say, on some region around x^* all of the function values are greater than or equal to the value at that point. Local maxima are defined similarly.

A large number of algorithms proposed for solving non-convex problems – including the majority of commercially available solvers

– are not capable of making a distinction between local optimal solutions and rigorous optimal solutions, and will treat the former as actual solutions to the original problem.

The branch of applied mathematics and numerical analysis that is concerned with the development of deterministic algorithms that are capable of guaranteeing convergence in finite time to the actual optimal solution of a non-convex problem is called global optimisation.

Notation

Optimisation problems are often expressed with special notation. Here are some examples.

Minimum and Maximum Value of a Function

Consider the following notation:

$$\min_{x \in \mathbb{R}} (x^2 + 1)$$

This denotes the minimum value of the objective function $x^2 + 1$, when choosing x from the set of real numbers $\mathbb{R}$. The minimum value in this case is 1, occurring at $x = 0$.

Similarly, the notation

$$\max_{x \in \mathbb{R}} 2x$$

asks for the maximum value of the objective function $2x$, where x may be any real number. In this case, there is no such maximum as the objective function is unbounded, so the answer is "infinity" or "undefined".

Optimal Input Arguments

Consider the following notation:

$$\underset{x \in (-\infty, -1]}{\operatorname{argmin}} \; x^2 + 1,$$

or equivalently

$$\underset{x}{\operatorname{argmin}} \; x^2 + 1, \text{subject to: } x \in (-\infty, -1].$$

This represents the value (or values) of the argument x in the interval $(-\infty, -1]$ that minimises (or minimise) the objective function $x^2 + 1$ (the actual minimum value of that function is not what the problem asks for). In this case, the answer is $x = -1$, since $x = 0$ is infeasible, i.e. does not belong to the feasible set.

Similarly,

$$\underset{x\in[-5,5],\ y\in\mathbb{R}}{\operatorname{argmax}}\ x\cos(y),$$

or equivalently

$$\operatorname{argmax}\ x\cos(y),\ \text{subject to: } x\in[-5,5],\ y\in\mathbb{R},$$

represents the (x,y) pair (or pairs) that maximises (or maximise) the value of the objective function $x\cos(y)$, with the added constraint that x lie in the interval $[-5,5]$ (again, the actual maximum value of the expression does not matter). In this case, the solutions are the pairs of the form $(5, 2k\pi)$ and $(-5,(2k+1)\pi)$, where k ranges over all integers.

Argmin and argmax are sometimes also written argmin and argmax, and stand for argument of the minimum and argument of the maximum.

History

Fermat and Lagrange found calculus-based formulas for identifying optima, while Newton and Gauss proposed iterative methods for moving towards an optimum. Historically, the first term for optimisation was "linear programming", which was due to George B. Dantzig, although much of the theory had been introduced by Leonid Kantorovich in 1939. Dantzig published the Simplex algorithm in 1947, and John von Neumann developed the theory of duality in the same year.

The term *programming* in this context does not refer to computer programming. Rather, the term comes from the use of *program* by the United States military to refer to proposed training and logistics schedules, which were the problems Dantzig studied at that time.

Later important researchers in mathematical optimisation include the following:

- Richard Bellman
- Ronald A. Howard
- Narendra Karmarkar
- William Karush
- Leonid Khachiyan
- Bernard Koopman
- Harold Kuhn

- Joseph Louis Lagrange
- Laszlo Lovasz
- Arkadii Nemirovskii
- Yurii Nesterov
- Boris Polyak
- Lev Pontryagin
- James Renegar
- R. Tyrrell Rockafellar
- Cornelis Roos
- Naum Z. Shor
- Michael J. Todd
- Albert Tucker.

Major subfields:

- Convex programming studies the case when the objective function is convex (minimisation) or concave (maximisation) and the constraint set is convex. This can be viewed as a particular case of nonlinear programming or as generalisation of linear or convex quadratic programming.
- Linear programming (LP), a type of convex programming, studies the case in which the objective function *f* is linear and the set of constraints is specified using only linear equalities and inequalities. Such a set is called a polyhedron or a polytope if it is bounded.
- Second order cone programming (SOCP) is a convex program, and includes certain types of quadratic programs.
- Semidefinite programming (SDP) is a subfield of convex optimisation where the underlying variables are semidefinite matrices. It is generalisation of linear and convex quadratic programming.
- Conic programming is a general form of convex programming. LP, SOCP and SDP can all be viewed as conic programs with the appropriate type of cone.
- Geometric programming is a technique whereby objective and inequality constraints expressed as posynomials and equality constraints as monomials can be transformed into a convex program.

- Integer programming studies linear programs in which some or all variables are constrained to take on integer values. This is not convex, and in general much more difficult than regular linear programming.
- Quadratic programming allows the objective function to have quadratic terms, while the feasible set must be specified with linear equalities and inequalities. For specific forms of the quadratic term, this is a type of convex programming.
- Fractional programming studies optimisation of ratios of two nonlinear functions. The special class of concave fractional programs can be transformed to a convex optimisation problem.
- Nonlinear programming studies the general case in which the objective function or the constraints or both contain nonlinear parts. This may or may not be a convex program. In general, whether the program is convex affects the difficulty of solving it.
- Stochastic programming studies the case in which some of the constraints or parameters depend on random variables.
- Robust programming is, like stochastic programming, an attempt to capture uncertainty in the data underlying the optimisation problem. This is not done through the use of random variables, but instead, the problem is solved taking into account inaccuracies in the input data.
- Combinatorial optimisation is concerned with problems where the set of feasible solutions is discrete or can be reduced to a discrete one.
- Infinite-dimensional optimisation studies the case when the set of feasible solutions is a subset of an infinite-dimensional space, such as a space of functions.
- Metaheuristics make few or no assumptions about the problem being optimised and can search very large spaces of candidate solutions. However, metaheuristics do not guarantee an optimal solution is ever found.
- Constraint satisfaction studies the case in which the objective function f is constant (this is used in artificial intelligence, particularly in automated reasoning).
- Constraint programming.
- Disjunctive programming is used where at least one constraint must be satisfied but not all. It is of particular use in scheduling.

In a number of subfields, the techniques are designed primarily for optimisation in dynamic contexts (that is, decision making over time):

- Calculus of variations seeks to optimise an objective defined over many points in time, by considering how the objective function changes if there is a small change in the choice path.
- Optimal control theory is a generalisation of the calculus of variations.
- Dynamic programming studies the case in which the optimisation strategy is based on splitting the problem into smaller subproblems. The equation that describes the relationship between these subproblems is called the Bellman equation.
- Mathematical programming with equilibrium constraints is where the constraints include variational inequalities or complementarities.

Multi-objective Optimisation

Adding more than one objective to an optimisation problem adds complexity. For example, to optimise a structural design, one would want a design that is both light and rigid. Because these two objectives conflict, a trade-off exists. There will be one lightest design, one stiffest design, and an infinite number of designs that are some compromise of weight and stiffness. The set of trade-off designs that cannot be improved upon according to one criterion without hurting another criterion is known as the Pare to set. The curve created plotting weight against stiffness of the best designs is known as the Pare to frontier.

A design is judged to be "Pare to optimal" (equivalently, "Pare to efficient" or in the Pare to set) if it is not dominated by any other design: If it is worse than another design in some respects and no better in any respect, then it is dominated and is not Pare to optimal.

Multi-modal Optimisation

Optimisation problems are often multi-modal; that is they possess multiple good solutions. They could all be globally good (same cost function value) or there could be a mix of globally good and locally good solutions. Obtaining all (or at least some of) the multiple solutions is the goal of a multi-modal optimiser. Classical optimisation techniques due to their iterative approach do not perform satisfactorily when they are used to obtain multiple solutions, since it is not guaranteed that different solutions will be obtained even with different starting points in multiple runs of the algorithm. Evolutionary Algorithmsare however a very popular approach to obtain multiple solutions in a multi-modal optimisation task.

Classification of Critical Points and Extrema

Feasibility Problem

The satisfiability problem, also called the feasibility problem, is just the problem of finding any feasible solution at all without regard to objective value. This can be regarded as the special case of mathematical optimisation where the objective value is the same for every solution, and thus any solution is optimal.

Many optimisation algorithms need to start from a feasible point. One way to obtain such a point is to relax the feasibility conditions using a slack variable; with enough slack, any starting point is feasible. Then, minimise that slack variable until slack is null or negative.

Existence

The extreme value theorem of Karl Weierstrass states that a continuous real-valued function on a compact set attains its maximum and minimum value. More generally, a lower semi-continuous function on a compact set attains its minimum; an upper semi-continuous function on a compact set attains its maximum.

Necessary Conditions for Optimality

One of Fermat's theorems states that optima of unconstrained problems are found at stationary points, where the first derivative or the gradient of the objective function is zero. More generally, they may be found at critical points, where the first derivative or gradient of the objective function is zero or is undefined, or on the boundary of the choice set.

An equation (or set of equations) stating that the first derivative(s) equal(s) zero at an interior optimum is called a 'first-order condition' or a set of first-order conditions.

Optima of inequality-constrained problems are instead found by the Lagrange multiplier method. This method calculates a system of inequalities called the 'Karush–Kuhn–Tucker conditions' or 'complementary slackness conditions', which may then be used to calculate the optimum.

Sufficient Conditions for Optimality

While the first derivative test identifies points that might be optima, this test does not distinguish a point which is a minimum from one that is a maximum or one that is neither. When the objective function is twice differentiable, these cases can be distinguished by checking the second derivative or the matrix of second derivatives (called the Hessian matrix) in unconstrained problems, or the matrix of second derivatives of the objective function and the constraints called the bordered Hessian in constrained problems.

The conditions that distinguish maxima, or minima, from other stationary points are called 'second-order conditions'. If a candidate solution satisfies the first-order conditions, then satisfaction of the second-order conditions as well is sufficient to establish at least local optimality.

Sensitivity and Continuity of Optima

The envelope theorem describes how the value of an optimal solution changes when an underlying parameter changes. The process of computing this change is called comparative statics.

The maximum theorem of Claude Berge (1963) describes the continuity of an optimal solution as a function of underlying parameters.

Calculus of Optimisation

For unconstrained problems with twice-differentiable functions, some critical points can be found by finding the points where the gradient of the objective function is zero (that is, the stationary points). More generally, a zero subgradient certifies that a local minimum has been found for minimisation problems with convex functions and other locally Lipschitz functions.

Further, critical points can be classified using the definiteness of the Hessian matrix: If the Hessian is positive definite at a critical point, then the point is a local minimum; if the Hessian matrix is negative definite, then the point is a local maximum; finally, if indefinite, then the point is some kind of saddle point. Constrained problems can often be transformed into unconstrained problems with the help of Lagrange multipliers. Lagrangian relaxation can also provide approximate solutions to difficult constrained problems.

When the objective function is convex, then any local minimum will also be a global minimum. There exist efficient numerical techniques for minimising convex functions, such as interior-point methods.

Computational Optimisation Techniques

To solve problems, researchers may use algorithms that terminate in a finite number of steps, oriterative methods that converge to a solution (on some specified class of problems), or heuristics that may provide approximate solutions to some problems (although their iterates need not converge).

Optimisation Algorithms

- Simplex algorithm of George Dantzig, designed for linear programming.
- Extensions of the simplex algorithm, designed for quadratic programming and for linear-fractional programming.
- Variants of the simplex algorithm that are especially suited for network optimisation.
- Combinatorial algorithms.

Iterative Methods

The iterative methods used to solve problems of nonlinear programming differ according to whether they evaluate Hessians, gradients, or only function values. While evaluating Hessians (H) and gradients (G) improves the rate of convergence, such evaluations increase the computational complexity (or computational cost) of each iteration. In some cases, the computational complexity may be excessively high. One major criterion for optimisers is just the number of required function evaluations as this often is already a large computational effort, usually much more effort than within the optimiser itself, which mainly has to operate over the N variables. The derivatives provide detailed information for such optimisers, but are even harder to calculate, e.g. approximating the gradient takes at least N+1 function evaluations. For approximations of the 2nd derivatives (collected in the Hessian matrix) the number of function evaluations is in the order of N^2. Newton's method requires the 2nd order derivates, so for each iteration the number of function calls is in the order of N^2, but for a simpler pure gradient optimiser it is only N. However, gradient optimisers need usually more iterations than Newton's algorithm. Which one is best wrt. Number of function calls depends on the problem itself.

- Methods that evaluate Hessians (or approximate Hessians, using finite differences):
- Newton's method
- Sequential quadratic programming: A method for small-medium scale constrained problems. Some versions can handle large-dimensional problems.
- Methods that evaluate gradients or approximate gradients using finite differences (or even subgradients):
- Quasi-Newton methods: Iterative methods for medium-large problems (e.g. N<1000).
- Conjugate gradient methods: Iterative methods for large problems. (In theory, these methods terminate in a finite number of steps with quadratic objective functions, but this finite termination is not observed in practice on finite–precision computers.)
- Interior point methods: This is a large class of methods for constrained optimisation. Some interior-point methods use only (sub)gradient information, and others of which require the evaluation of Hessians.
- Gradient descent (alternatively, "steepest descent" or "steepest ascent"): A (slow) method of historical and theoretical interest, which has had renewed interest for finding approximate solutions of enormous problems.
- Subgradient methods - An iterative method for large locally Lipschitz functions using generalised gradients. Following Boris T. Polyak, subgradient–projection methods are similar to conjugate–gradient methods.
- Bundle method of descent: An iterative method for small–medium sized problems with locally Lipschitz functions, particularly for convex minimisation problems. (Similar to conjugate gradient methods)
- Ellipsoid method: An iterative method for small problems with quasiconvex objective functions and of great theoretical interest, particularly in establishing the polynomial time complexity of some combinatorial optimisation problems. It has similarities with Quasi-Newton methods.
- Reduced gradient method (Frank–Wolfe) for approximate minimisation of specially structured problems with linear

constraints, especially with traffic networks. For general unconstrained problems, this method reduces to the gradient method, which is regarded as obsolete (for almost all problems).

- Methods that evaluate only function values: If a problem is continuously differentiable, then gradients can be approximated using finite differences, in which case a gradient-based method can be used. Usually such non-gradient methods work only fine for problems with few parameter only.
- Interpolation methods
- Pattern search methods, which have better convergence properties than the Nelder–Mead simplicial heuristic (listed below, under heuristics).

Global Convergence

More generally, if the objective function is not a quadratic function, then many optimisation methods use other methods to ensure that some subsequence of iterations converges to an optimal solution. The first and still popular method for ensuring convergence relies on-line searches, which optimise a function along one dimension. A second and increasingly popular method for ensuring convergence uses trust regions. Both line searches and trust regions are used in modern methods of non-differentiable optimisation. Usually a global optimiser is much slower than advanced local optimisers (such as BFGS), so often an efficient global optimiser can be constructed by starting the local optimiser from different starting points.

Heuristics

Besides (finitely terminating) algorithms and (convergent) iterative methods, there are heuristics that can provide approximate solutions to some optimisation problems:

- Memetic algorithm
- Differential evolution
- Dynamic relaxation
- Genetic algorithms
- Hill climbing
- Nelder-Mead simplicial heuristic: A popular heuristic for approximate minimisation (without calling gradients)
- Particle swarm optimisation

- Simulated annealing
- Tabu search

Applications

Mechanics and Engineering

Problems in rigid body dynamics (in particular articulated rigid body dynamics) often require mathematical programming techniques, since you can view rigid body dynamics as attempting to solve an ordinary differential equation on a constraint manifold; the constraints are various nonlinear geometric constraints such as "these two points must always coincide", "this surface must not penetrate any other", or "this point must always lie somewhere on this curve". Also, the problem of computing contact forces can be done by solving a linear complementarity problem, which can also be viewed as a QP (quadratic programming) problem.

Many design problems can also be expressed as optimisation programs. This application is called design optimisation. One subset is the engineering optimisation, and another recent and growing subset of this field is multidisciplinary design optimisation, which, while useful in many problems, has in particular been applied to aerospace engineering problems.

Economics

Economics is closely enough linked to optimisation of agents that an influential definition relatedly describes economics *qua* science as the "study of human behaviour as a relationship between ends and scarce means" with alternative uses. Modern optimisation theory includes traditional optimisation theory but also overlaps with game theory and the study of economic equilibria. The *Journal of Economic Literature* codes classify mathematical programming, optimisation techniques, and related topics under JEL:C61-C63.

In microeconomics, the utility maximisation problem and its dual problem, the expenditure minimisation problem, are economic optimisation problems. Insofar as they behave consistently, consumers are assumed to maximise their utility, while firms are usually assumed to maximise their profit. Also, agents are often modelled as being risk-averse, thereby preferring to avoid risk. Asset prices are also modelled using optimisation theory, though the underlying mathematics relies on optimising stochastic processes rather than on static optimisation. Trade theory also uses optimisation to explain trade patterns between nations. Since the 1970s, economists have modelled dynamic decisions over time using control theory. For example, microeconomists use dynamic search models to study labour-market behaviour. A crucial distinction is between deterministic and stochastic models. Macroeconomists build dynamic stochastic general equilibrium (DSGE) models that describe the dynamics of the whole economy as the result of the interdependent optimising decisions of workers, consumers, investors, and governments.

Operations Research

Another field that uses optimisation techniques extensively is operations research. Operations research also uses stochastic modelling and simulation to support improved decision-making. Increasingly, operations research uses stochastic programming to model dynamic decisions that adapt to events; such problems can be solved with large-scale optimisation and stochastic optimisation methods.

Metaheuristic

In computer science, metaheuristic designates a computational method that optimises a problem byiteratively trying to improve a candidate solution with regard to a given measure of quality. Metaheuristics make few or no assumptions about the problem being optimised and can search very large spaces of candidate solutions. However, metaheuristics do not guarantee an optimal solution is ever found. Many metaheuristics implement some form of stochastic optimisation.

Other terms having a similar meaning as metaheuristic, are: derivative-free, direct search, black-box, or indeed just heuristic optimiser. Several books and survey papers have been published on the subject.

Applications

Metaheuristics are used for combinatorial optimisation in which an optimal solution is sought over a discrete search-space. An example problem is the travelling salesman problem where the search-space of candidate solutions grows more than exponentially as the size of the problem increases, which makes an exhaustive search for the optimal solution infeasible. Additionally, multidimensional combinatorial problems, including most design problems in engineering such as form-finding and behaviour-finding, suffer from the curse of dimensionality, which also akes them infeasible for exhaustive search or analytical methods. Popular metaheuristics for combinatorial problems include simulated annealing by Kirkpatrick et al., genetic algorithms by Holland et al., ant colony optimisation by Dorigo, scatter search and tabu search by Glover.

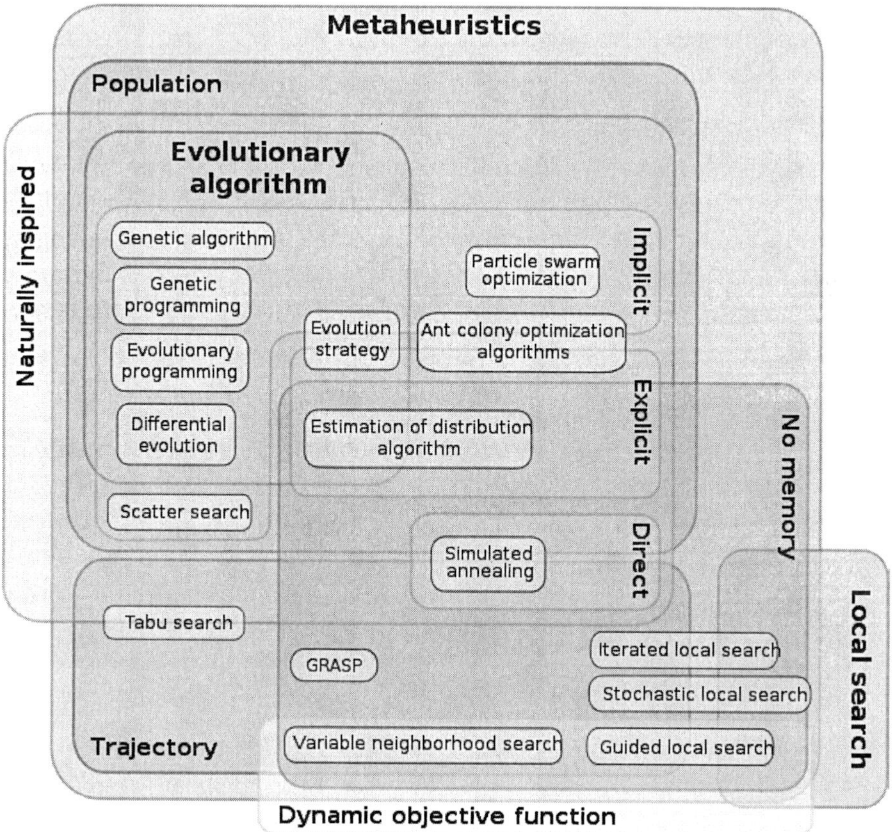

Figure: Different classifications of metaheuristics.

Metaheuristics are also used for problems over real-valued search-spaces, where the classic way of optimisation is to derive the gradient of the function to be optimised and then employ gradient descentor a quasi-Newton method. Metaheuristics do not use the gradient or Hessian matrix so their advantage is that the function to be optimised need not be continuous or differentiable and it can also have constraints. Popular metaheuristic optimisers for real-valued search-spaces include particle swarm optimisation by Eberhart and Kennedy, differential evolution by Storn and Price and evolution strategies by Rechenberg and Schwefel.

Metaheuristics based on decomposition techniques have also been proposed for tackling hard combinatorial problems of large size.

Main Contributions

Many different metaheuristics are in existence and new variants are continually being proposed. Some of the most significant contributions to the field are:

- 1952: Robbins and Monro work on stochastic optimisation methods.
- 1952: Fermi and Metropolis develop an early form of pattern search as described belatedly by Davidon.
- 1954: Barricelli carry out the first simulations of the evolution process and use them on general optimisation problems.
- 1963: Rastrigin proposes random search.
- 1965: Matyas proposes random optimisation.
- 1965: Rechenberg proposes evolution strategies.
- 1965: Nelder and Mead propose a simplex heuristic, which was shown by Powell to converge to non-stationary points on some problems.
- 1966: Fogel et al. propose evolutionary programming.
- 1970: Hastings proposes the Metropolis-Hastings algorithm.
- 1970: Cavicchio proposes adaptation of control parameters for an optimiser.
- 1970: Kernighan and Lin propose a graph partitioning method, related to variable-depth search and prohibition-based (tabu) search.
- 1975: Holland proposes the genetic algorithm.
- 1977: Glover proposes Scatter Search.
- 1978: Mercer and Sampson propose a metaplan for tuning an optimiser's parameters by using another optimiser.
- 1980: Smith describes genetic programming.
- 1983: Kirkpatrick et al. propose simulated annealing.
- 1986: Glover proposes tabu search, first mention of the term *metaheuristic*.
- 1986: Farmer et al. work on the artificial immune system.
- 1986: Grefenstette proposes another early form of metaplan for tuning an optimiser's parameters by using another optimiser.

- 1988: First conference on genetic algorithms is organised at the University of Illinois at Urbana-Champaign.
- 1988: Koza registers his first patent on genetic programming.
- 1989: Goldberg publishes a well known book on genetic algorithms.
- 1989: Evolver, the first optimisation software using the genetic algorithm.
- 1989: Moscato proposes the memetic algorithm.
- 1991: Interactive evolutionary computation.
- 1992: Dorigo proposes the ant colony algorithm.
- 1993: The journal, *Evolutionary Computation*, begins publication by the Massachusetts Institute of Technology.
- 1993: Fonseca and Fleming propose MOGA for multiobjective optimisation.
- 1994: Battiti and Tecchiolli propose Reactive Search Optimisation(RSO) principles for the online self-tuning of heuristics.
- 1994: Srinivas and Deb propose NSGA for multiobjective optimisation.
- 1995: Kennedy and Eberhart propose particle swarm optimisation.
- 1995: Wolpert and Macready prove the no free lunch theorems.
- 1996: Mühlenbein and Paaß work on the estimation of distribution algorithm.
- 1996: Hansen and Ostermeier propose CMA-ES.
- 1997: Storn and Price propose differential evolution.
- 1997: Rubinstein proposes the cross entropy method.
- 1999: Taillard and Voss propose POPMUSIC.
- 2001: Geem et al. propose harmony search.
- 2002: Deb et al. propose NSGA-II for multiobjective optimisation.
- 2004: Nakrani and Tovey propose bee colony optimisation.
- 2005: Krishnanand and Ghose propose Glowworm swarm optimisation.
- 2005: Karaboga proposes Artificial Bee Colony Algorithm (ABC).

- 2006: Haddad et al. introduces honey-bee mating optimisation.
- 2007: Hamed Shah-Hosseini introduces Intelligent Water Drops.
- 2008: Wierstra et al. propose natural evolution strategies based on the natural gradient.
- 2008: Yang introduces firefly algorithm.
- 2008: Mucherino and Seref propose the Monkey Search
- 2009: Yang and Deb introduce cuckoo search.
- 2011: Hamed Shah-Hosseini proposes the Galaxy-based Search Algorithm.
- 2011: Tamura and Yasuda propose spiral optimisation.

Criticism

Mathematical analyses of metaheuristics have been presented in the literature, e.g. Holland's schema theorem for the genetic algorithm, the work by Trelea, amongst others, for analysis of particle swarm optimisation, and Zaharie for analysis of differential evolution. These analyses make a number of assumptions in regard to the optimiser variants and the simplicity of the optimisation problems which limit their validity in real-world optimisation scenarios. Performance and convergence aspects of metaheuristic optimisers are therefore often demonstrated empirically in the research literature. This has been criticized in the no free lunch set of theorems by Wolpert and Macready, which, among other things, prove that all optimisers perform equally well when averaged over all problems. The practical relevance of the no free lunch theorems however is minor, because they will generally not hold on the collection of problems a practitioner is facing.

For the practitioner the most relevant issue is that metaheuristics are not guaranteed to find the optimum or even a satisfactory near-optimal solution. All metaheuristics will eventually encounter problems on which they perform poorly and the practitioner must gain experience in which optimisers work well on different classes of problems.

Function of a Real Variable

In mathematics, a function of a real variable is a mathematical function whose domain is the real line. More loosely, a function of a real variable is sometimes taken to mean any function whose domain is a subset of the real line. Functions of a real variable were the classical object of study in mathematical analysis, specifically real analysis. In that context, a function of a real variable was usually meant a real-valued function of a real variable, that is, a function whose domain and codomain were the real numbers. However, because of their convenience in fields such as Fourier analysis, it was also common to consider complex functions of a real variable, that is, a function whose domain was the real numbers and whose range was the complex numbers.

Until the introduction of functional analysis in the 1920s, the study of functions of a real variable was one of the two major subdivisions of mathematical analysis. The other was the study of functions of a complex variable, otherwise known as complex analysis.

Particle Swarm Optimisation

In computer science, particle swarm optimisation (PSO) is a computational method that optimises a problem by iteratively trying to improve a candidate solution with regard to a given measure of quality. PSO optimises a problem by having a population of candidate solutions, here dubbed particles, and moving these particles around in the search-space according to simple mathematical formulae over the particle's position and velocity.

Each particle's movement is influenced by its local best known position and is also guided toward the best known positions in the search-space, which are updated as better positions are found by other particles. This is expected to move the swarm toward the best solutions.

PSO is originally attributed to Kennedy, Eberhart and Shi and was first intended for simulating social behaviour, as a stylized representation of the movement of organisms in a bird flock or fish school. The algorithm was simplified and it was observed to be performing optimisation. The book by Kennedy and Eberhart describes many philosophical aspects of PSO and swarm intelligence. An extensive survey of PSO applications is made by Poli.

PSO is a metaheuristic as it makes few or no assumptions about the problem being optimised and can search very large spaces of candidate solutions. However, metaheuristics such as PSO do not guarantee an optimal solution is ever found. More specifically, PSO does not use the gradient of the problem being optimised, which means PSO does not require that the optimisation problem be differentiable as is required by classic optimisation methods such as gradient descent and quasi-newton methods. PSO can therefore also be used on optimisation problems that are partially irregular, noisy, change over time, etc.

Algorithm

A basic variant of the PSO algorithm works by having a population (called a swarm) of candidate solutions (called particles). These particles are moved around in the search-space according to a few simple formulae.

The movements of the particles are guided by their own best known position in the search-space as well as the entire swarm's best known position. When improved positions are being discovered these will then come to guide the movements of the swarm. The process is repeated and by doing so it is hoped, but not guaranteed, that a satisfactory solution will eventually be discovered.

Formally, let $f: \mathbb{R}^n \rightarrow \mathbb{R}$ be the fitness or cost function which must be minimised. The function takes a candidate solution as argument in the form of a vector of real numbers and produces a real number as output which indicates the fitness of the given candidate solution. The gradient of f is not known. The goal is to find a solution a for which f(a) d– f(b) for all b in the search-space, which would mean a is the global minimum. Maximisation can be performed by considering the function $h = -f$ instead.

Let S be the number of particles in the swarm, each having a position $x_i \in \mathbb{R}^n$ in the search-space and a velocity $v_i \in \mathbb{R}^n$. Let p_i be the best known position of particle i and let g be the best known position of the entire swarm. A basic PSO algorithm is then:

- For each particle $i = 1, ..., S$ do:
- Initialize the particle's position with a uniformly distributed random vector: $x_i \sim U(b_{lo}, b_{up})$, where b_{lo} and b_{up} are the lower and upper boundaries of the search-space.
- Initialize the particle's best known position to its initial position: $p_i \leftarrow x_i$
- If $(f(p_i) < f(g))$ update the swarm's best known position: $g \leftarrow p_i$
- Initialize the particle's velocity: $v_i \sim U(-|b_{up}-b_{lo}|, |b_{up}-b_{lo}|)$
- Until a termination criterion is met (e.g. number of iterations performed, or adequate fitness reached), repeat:
- For each particle $i = 1, ..., S$ do:
- Pick random numbers: $r_p, r_g \sim U(0,1)$

- Update the particle's velocity: $v_i \leftarrow \omega\, v_i + \varphi_p\, r_p\, (p_i\text{-}x_i) + \varphi_g\, r_g\, (g\text{-}x_i)$
- Update the particle's position: $x_i \leftarrow x_i + v_i$
- If $(f(x_i) < f(p_i))$ do:
- Update the particle's best known position: $p_i \leftarrow x_i$
- If $(f(p_i) < f(g))$ update the swarm's best known position: $g \leftarrow p_i$
- Now g holds the best found solution.

The parameters ω, φ_p, and φ_g are selected by the practitioner and control the behaviour and efficacy of the PSO method.

Parameter Selection

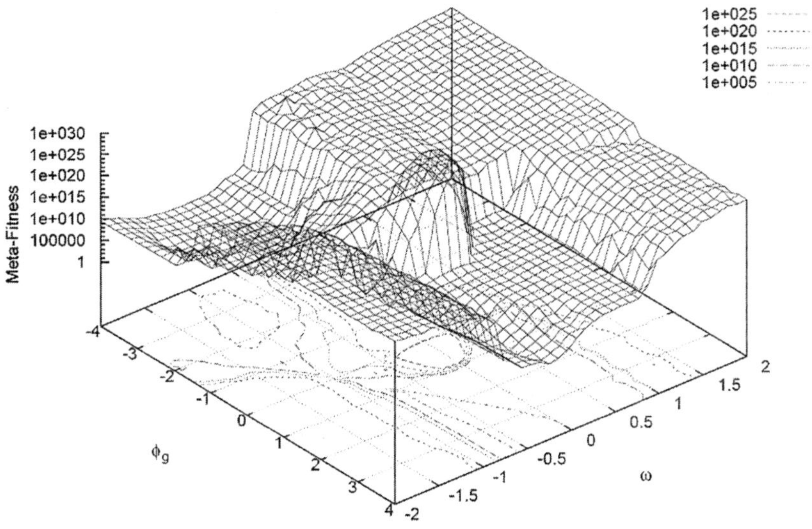

Figure: Performance landscape showing how a simple PSO variant performs in aggregate on several benchmark problems when varying two PSO parameters.

The choice of PSO parameters can have a large impact on optimisation performance. Selecting PSO parameters that yield good performance has therefore been the subject of much research. The PSO parameters can also be tuned by using another overlaying optimiser, a concept known as meta-optimisation. Parameters have also been tuned for various optimisation scenarios.

Inner Workings

There are several schools of thought as to why and how the PSO algorithm can perform optimisation.

A common belief amongst researchers is that the swarm behaviour varies between exploratory behaviour, that is, searching a broader region of the search-space, and exploitative behaviour, that is, a locally oriented search so as to get closer to a (possibly local) optimum. This school of thought has been prevalent since the inception of PSO. This school of thought contends that the PSO algorithm and its parameters must be chosen so as to properly balance between exploration and exploitation to avoid premature convergence to a local optimum yet still ensure a good rate of convergence to the optimum. This belief is the precursor of many PSO variants.

Another school of thought is that the behaviour of a PSO swarm is not well understood in terms of how it affects actual optimisation performance, especially for higher dimensional search-spaces and optimisation problems that may be discontinuous, noisy, and time-varying. This school of thought merely tries to find PSO algorithms and parameters that cause good performance regardless of how the swarm behaviour can be interpreted in relation to e.g. exploration and exploitation. Such studies have led to the simplification of the PSO algorithm.

Convergence

In relation to PSO the word *convergence* typically means one of two things, although it is often not clarified which definition is meant and sometimes they are mistakenly thought to be identical.

- Convergence may refer to the swarm's best known position g approaching (converging to) the optimum of the problem, regardless of how the swarm behaves.
- Convergence may refer to a swarm collapse in which all particles have converged to a point in the search-space, which may or may not be the optimum.

Several attempts at mathematically analysing PSO convergence exist in the literature. These analyses have resulted in guidelines for selecting PSO parameters that are believed to cause convergence, divergence or oscillation of the swarm's particles, and the analyses have also given rise to several PSO variants. However, the analyses were criticized by Pedersen for being oversimplified as they assume the swarm has only one particle, that it does not use stochastic variables and that the points of attraction, that is, the particle's best known position p and the swarm's best known position g, remain constant throughout the optimisation process. Furthermore, some analyses allow for an infinite number of optimisation iterations which is not possible in reality. This means that determining convergence capabilities of different PSO algorithms and parameters therefore still depends on empirical results.

Variants

Numerous variants of even a basic PSO algorithm are possible. For example, there are different ways to initialize the particles and velocities (e.g. start with zero velocities instead), how to dampen the velocity, pick random numbers r_p and r_g for each dimension when updating the velocity-vector rather than using the same numbers for the entire vector, only update p_i and g after the entire swarm has been updated, use a subset of the swarm (known as a neighbourhood) instead of g when updating a particle's velocity, etc. Some of these choices and their possible performance impact have been discussed in the literature.

New and more sophisticated PSO variants are also continually being introduced in an attempt to improve optimisation performance. There are certain trends in that research; one is to make a hybrid optimisation method using PSO combined with other optimisers , another research trend is to try and alleviate premature convergence (that is, optimisation stagnation) e.g. by reversing or perturbing the movement of the PSO particles , another approach to deal with premature convergence is the use of multiple swarms (multi-swarm optimisation), and then there are also attempts at adapting the behavioural parameters of PSO during optimisation.

Simplifications

Another school of thought is that PSO should be simplified as much as possible without impairing its performance; a general concept often referred to as Occam's razor. Simplifying PSO was originally suggested by Kennedy and has been studied more extensively, where it appeared that optimisation performance was improved, and the parameters were easier to tune and they performed more consistently across different optimisation problems.

Another argument in favour of simplifying PSO is that metaheuristics cannly have their efficacy emonstrated empirically by doing computational experiments on a finite number of optimisation problems. This means a metaheuristic such as PSO cannot be proven correct and this increases the risk of making errors in its description and implementation. A good example of this presented a promising variant of a genetic algorithm (another popular metaheuristic) but it was later found to be defective as it was strongly biased in its optimisation search towards similar values for different dimensions in the search space, which happened to be the optimum of the benchmark problems considered. This bias was because of a programming error, and has now been fixed.

Multi-objective Optimisation

PSO has also been applied to multi-objective problems, in which the fitness comparison takes pare to dominance into account when moving the PSO particles and non-dominated solutions are stored so as to approximate the pare to front.

Chapter 2

Important Terms and Persons

James Kennedy (Social Psychologist)

James Kennedy (b. November 5, 1950) is an American social psychologist, best known as an originator and researcher of particle swarm optimization. The first papers on the topic, by Kennedy andRussell C. Eberhart, were presented in 1995; since then more than ten thousand papers have been published on particle swarms. The Academic Press / Morgan Kaufmann book, *Swarm Intelligence*, by Kennedy and Eberhart with Yuhui Shi, was published in 2001.

The particle swarm paradigm draws on social-psychological simulation research in which Kennedy had participated at the University of North Carolina, integrated with evolutionary computation methods that Eberhart had been working with in the 1990s. The result was a problem-solving or optimization algorithm based on the principles of human social interaction. Individuals begin the program with random guesses at the problem solution. As the program runs, the "particles" share their successes with their topological neighbors; each particle is both teacher and learner. Over time, the population converges reliably on optimal vectors.

Kennedy is an active combatant in the controversy over sex education in Montgomery County, Maryland, supporting the public schools' efforts to develop a comprehensive and inclusive program. He also worked to support a gender identity nondiscrimination law in Montgomery County that came under attack from conservatives. He also worked as a professional musician for twenty years and currently plays in a rockabilly band called The Colliders.

According to the back cover of his 2001 *Swarm Intelligence* book, Kennedy works in survey methods at the US Department of Labor, has conducted basic and applied research into the social effects oncognition and attitude. He has worked with particle swarms since 1994.

Social Behavior

In physics, physiology and sociology, social behavior is behavior directed towards society, or taking place between, members of the same species. Behavior such as predation which involves members of different species is not social. While many social behaviors are communication (provoke a response, or change in behavior, without acting directly on the receiver) communication between members of different species is not social behavior. The umbrella term behavioral sciences is used to refer to sciences that study behaviorality disturbance in general.

In sociology, "behavior" itself means an animal-like activity devoid of social meaning or social context, in contrast to "social behavior" which has both. In a sociological hierarchy, social behavior is followed by social actions, which is directed at other people and is designed to induce a response. Further along this ascending scale are social interaction and social relation. In conclusion, social behavior is a process of communicating.

Among specific social behaviors are regarded, e.g., aggression, altruism, scapegoating andshyness.

'Monosociality' describes social relations (or preference for such relations) with the same sex of a (putatively) nonsexual nature. 'Bisociality' describes social relations (or preference for such relations) with both the same and opposite sexes, also of a (putatively) nonsexual nature.

Flocking (Behavior)

Flocking behavior is the behavior exhibited when a group of birds, called a flock, are foraging or in flight. There are parallels with the shoaling behavior of fish, the swarming behavior of insects, and herd behavior of land animals.

Computer simulations and mathematical models which have been developed to emulate the flocking behaviors of birds can generally be applied also to the "flocking" behavior of other species. As a result, the term "flocking" is sometimes applied, in computer science, to species other than birds. This article is about the modelling of flocking behavior. From the perspective of the mathematical modeller, "flocking" is the collective motion of a large number of self-propelled entities and is a collective animal behavior exhibited by many living beings such as birds, fish, bacteria, and insects. It is considered an emergent behavior arising from simple rules that are followed by individuals and does not involve any central coordination.

Flocking behavior was first simulated on a computer in 1986 by Craig Reynolds with his simulation program, Boids. This program simulates simple agents (boids) that are allowed to move according to a set of basic rules. The result is akin to a flock of birds, a school of fish, or a swarm of insects.

Flocking rules

Basic models of flocking behavior are controlled by three simple rules:

1. Separation - avoid crowding neighbors (short range repulsion)
2. Alignment - steer towards average heading of neighbors
3. Cohesion - steer towards average position of neighbors (long range attraction).

With these three simple rules, the flock moves in an extremely realistic way, creating complex motion and interaction that would be extremely hard to create otherwise.

The basic model has been extended in several different ways since Reynolds proposed it. For instance, Delgado-Mata et al. extended the basic model to incorporate the effects of fear. Olfaction was used to transmit emotion between animals, through pheromones modelled as particles in a free expansion gas. Hartman and Benes introduced a complementary force to the alignment that they call the change of leadership. This steer deûnes the chance of the boid to become a leader and try to escape. Hemerlijk and Hildenbrandt used attraction, alignment and avoidance and extended this with a numbet of traits of real starlings: first, birds fly according to fixed wing aerodynamics, while rolling when turning (thus losing lift), second they coordinate with a limited number of interaction neighbours of 7 (like in real starlings), third, they try to stay above a sleeping site (like starlings do at dawn) and when they happen to move outwards the sleeping site, they return to it by turning, fourth, they move at relative fixed speed. The authors showed that the specifics of flying behaviour as well as large flocksize and low number of interaction partners were essential to the creation of the variable shape of flocks of starlings.

Measurement

Measurements of bird flocking have been made using high-speed cameras, and a computer analysis has been made to test the simple rules of flocking mentioned above. It is found that they generally hold true in the case of bird flocking, but the long range attraction rule (cohesion) applies to the nearest 5-10 neighbors of the flocking bird and is independent of the distance of these neighbors from the bird. In addition, there is an anisotropy with regard to this cohesive tendency, with more cohesion being exhibited towards neighbors to the sides of the bird, rather than in front or behind. This is no doubt due to the field of vision of the flying bird being directed to the sides rather than directly forward or backward.

Algorithmic Complexity

In flocking simulations, there is no central control; each bird behaves autonomously. In other words, each bird has to decide for itself which flocks to consider as its environment. Usually environment is defined as a circle (2D) or sphere (3D) with a certain radius (representing reach).

A basic implementation of a flocking algorithm has complexity $O(n^2)$ - each bird searches through all other birds to find those who falls into his environment.

Possible improvements:

- bin-lattice spatial subdivision. Entire area the flock can move in is divided into a large number of bins. Each bin stores which birds it contains. Each time a bird moves from one bin to another, lattice has to be updated.
- Example: 2D(3D) grid in a 2D(3D) flocking simulation.
- Complexity: $O(nk)$, k is number of surrounding bins to consider; just when bird's bin is found in$O(1)$

Lee Spector, Jon Klein, Chris Perry and Mark Feinstein studied the emergence of collective behavior in evolutionary computation systems.

Applications

In Cologne, Germany, two biologists from the University of Leeds demonstrated a flock like behavior in humans. The group of people exhibited a very similar behavioral pattern to that of a flock, where if 5% of the flock would change direction the others would follow suit.

When one person was designated as a predator and everyone else was to avoid him, the flock behaved very much like a school of fish.

Flocking has also been considered as a means of controlling the be havior of Unmanned Air Vehicles (UAVs).

Flocking is a common technology in screensavers, and has found its use in animation. Flocking has been used in many films to generate crowds which move more realistically. Tim Burton's *Batman Returns* (1992) featured flocking bats, and Disney's *The Lion King* (1994) included a wildebeeststampede.

Flocking behaviour has been used for other interesting applications. It has been applied to automatically program Internet multi-channel radio stations . It has also been used for visualizing information and for optimization tasks .

Swarm Intelligence

Swarm intelligence (SI) is the collective behaviour of decentralized, self-organized systems, natural or artificial. The concept is employed in work on artificial intelligence. The expression was introduced by Gerardo Beni and Jing Wang in 1989, in the context of cellular robotic systems.

SI systems are typically made up of a population of simple agents or boids interacting locally with one another and with their environment. The inspiration often comes from nature, especially biological systems. The agents follow very simple rules, and although there is no centralized control structure dictating how individual agents should behave, local, and to a certain degree random, interactions between such agents lead to the emergence of "intelligent" global behavior, unknown to the individual agents. Natural examples of SI include ant colonies, bird flocking, animal herding, bacterial growth, and fish schooling.

The application of swarm principles to robots is called swarm robotics, while 'swarm intelligence' refers to the more general set of algorithms. 'Swarm prediction' has been used in the context of forecasting problems.

Example Algorithms

Altruism Algorithm

Researchers in Switzerland have developed an algorithm based on Hamilton's rule of kin selection. The algorithm shows how altruism in a swarm of entities can, over time, evolve and result in more effective swarm behaviour.

Ant Colony Optimization

Ant colony optimization (ACO) is a class of optimization algorithms modeled on the actions of an ant colony. ACO methods are useful in problems that need to find paths to goals. Artificial 'ants'—simulation agents—locate optimal solutions by moving through a parameter space representing all possible solutions. Real ants lay down pheromones directing each other to resources while exploring their environment. The simulated 'ants' similarly record their positions and the quality of their solutions, so that in later simulation iterations more ants locate better solutions. One variation on this approach is the bees algorithm, which is more analogous to the foraging patterns of the honey bee.

Artificial bee Colony Algorithm

Artificial Bee Colony (ABC) algorithm is a swarm based meta-heuristic algorithm introduced by Karaboga in 2005 , and simulates the foraging behaviour of honey bees.

The ABC algorithm has three phases: employed bee, onlooker bee and scout bee. In the employed bee and the onlooker bee phases, bees exploit the sources by local searches in the neighbourhood of the solutions selected based on deterministic selection in the employed bee phase and the probabilistic selection in the onlooker bee phase.

In the scout bee phase which is an analogy of abandoning exhausted food sources in the foraging process, solutions that are not beneficial anymore for search progress are abandoned, and new solutions are inserted instead of them to explore new regions in the search space. The algorithm has a good-balanced exploration and exploitation ability.

Artificial Immune Systems

Artificial immune systems (AIS) concerns the usage of abstract structure and function of the immune system to computational systems, and investigating the application of these systems towards solving computational problems from mathematics, engineering, and information technology. AIS is a sub-field of Biologically-inspired computing, and natural computation, with interests in Machine Learning and belonging to the broader field of Artificial Intelligence.

Charged System Search

Charged System Search (CSS) is a new optimization algorithm based on some principles from physics and mechanics. CSS utilizes the governing laws of Coulomb and Gauss from electrostatics and the Newtonian laws of mechanics. CSS is a multi-agent approach in which each agent is a Charged Particle (CP). CPs can affect each other based on their fitness values and their separation distances. The quantity of the resultant force is determined by using the electrostatics laws and the quality of the movement is determined using Newtonian mechanics laws. CSS is applicable to all optimization fields; especially it is suitable for non-smooth or non-convex domains. This algorithm provides a good balance between the exploration and the exploitation paradigms of the algorithm which can considerably improve the efficiency of the algorithm and therefore the CSS also can be considered as a good global and local optimizer simultaneously.

Cuckoo Search

Cuckoo search (CS) mimics the brooding behaviour of some cuckoo species, which use host birds to hatch their eggs and raise their chicks. This cuckoo search algorithm is enhanced with Levy flightswith jump steps drawn from Levy distribution. Recent studies suggested that CS can outperform other algorithms such as particle swarm optimization. For example, a comparison of the cuckoo search with PSO, DE and ABC suggest that CS and DE algorithms provide more robust results than PSO and ABC.

Firefly Algorithm

Firefly algorithm (FA) is another swarm-based algorithm, which was inspired by the flashing behaviour of fireflies. Light intensity is associated with attractiveness of a firefly, and such attraction enable the fireflies with the ability to subdivide into small groups and each subgroup swarm around the local modes. Therefore, firefly algorithm is specially suitable for multimodal optimization problems. In fact, FA has been applied in continuous optimization, traveling salesman problem, clustering, image processing and feature selection.

Gravitational Search Algorithm

Gravitational search algorithm (GSA) is constructed based on the law of gravity and the notion of mass interactions. The GSA algorithm uses the theory of Newtonian physics and its searcher agents are the collection of masses. In GSA, we have an isolated system of masses.

Using the gravitational force, every mass in the system can see the situation of other masses. The gravitational force is therefore a way of transferring information between different masses (Rashedi, Nezamabadi-pour and Saryazdi 2009). In GSA, agents are considered as objects and their performance is measured by their masses. All these objects attract each other by a gravity force, and this force causes a movement of all objects globally towards the objects with heavier masses. The heavy masses correspond to good solutions of the problem.

The position of the agent corresponds to a solution of the problem, and its mass is determined using a fitness function. By lapse of time, masses are attracted by the heaviest mass. We hope that this mass would present an optimum solution in the search space. The GSA could be considered as an isolated system of masses. It is like a small artificial world of masses obeying the Newtonian laws of gravitation and motion (Rashedi, Nezamabadi-pour and Saryazdi 2009). A multi-objective variant of GSA, called Non-dominated Sorting Gravitational Search Algorithm (NSGSA), was proposed by Nobahari and Nikusokhan in 2011.

Intelligent Water Drops

Intelligent Water Drops algorithm (IWD) is a swarm-based nature-inspired optimization algorithm, which has been inspired by natural rivers and how they find almost optimal paths to their destination. These near optimal or optimal paths follow from actions and reactions occurring among the water drops and the water drops with their riverbeds. In the IWD algorithm, several artificial water drops cooperate to change their environment in such a way that the optimal path is revealed as the one with the lowest soil on its links. The solutions are incrementally constructed by the IWD algorithm. Consequently, the IWD algorithm is generally a constructive population-based optimization algorithm.

Particle Swarm Optimization

Particle swarm optimization (PSO) is a global optimization algorithm for dealing with problems in which a best solution can be represented as a point or surface in an n-dimensional space. Hypotheses are plotted in this space and seeded with an initial velocity, as well as a communication channel between the particles. Particles then move through the solution space, and are evaluated according to some fitness criterion after each timestep. Over time, particles are accelerated towards those particles within their communication grouping which have better fitness values. The main advantage of such an approach over other global minimization strategies such as simulated annealingis that the large number of members that make up the particle swarm make the technique impressively resilient to the problem of local minima.

River Formation Dynamics

River formation dynamics (RFD) is an heuristic method similar to ant colony optimization (ACO). In fact, RFD can be seen as a gradient version of ACO, based on copying how water forms rivers by eroding the ground and depositing sediments. As water transforms the environment, altitudes of places are dynamically modified, and decreasing gradients are constructed.

The gradients are followed by subsequent drops to create new gradients, reinforcing the best ones. By doing so, good solutions are given in the form of decreasing altitudes. This method has been applied to solve different NP-complete problems (for example, the problems of finding a minimum distances tree and finding a minimum spanning tree in a variable-cost graph). The gradient orientation of RFD makes it specially suitable for solving these problems and provides a good tradeoff between finding good results and not spending much computational time. In fact, RFD fits particularly well for problems consisting in forming a kind of covering tree.

Self-propelled Particles

Self-propelled particles (SPP), also referred to as the *Couzin–Vicsek algorithm*, was introduced in 1995 by Vicsek and Couzin *et al.* as a special case of the boids model introduced in 1986 byReynolds. A swarm is modelled in SPP by a collection of particles that move with a constant speed but respond to a random perturbation by adopting at each time increment the average direction of motion of the other particles in their local neighbourhood. SPP models predict that swarming animals share certain properties at the group level, regardless of the type of animals in the swarm.Swarming systems give rise to emergent behaviours which occur at many different scales, some of which are turning out to be both universal and robust. It has become a challenge in theoretical physics to find minimal statistical models that capture these behaviours.

Stochastic Diffusion Search

Stochastic diffusion search (SDS) is an agent-based probabilistic global search and optimization technique best suited to problems where the objective function can be decomposed into multiple independent partial-functions. Each agent maintains a hypothesis which is iteratively tested by evaluating a randomly selected partial objective function parameterised by the agent's current hypothesis. In the standard version of SDS such partial function evaluations are binary, resulting in each agent becoming active or inactive. Information on hypotheses is diffused across the population via inter-agent communication. Unlike the stigmergic communication used in ACO, in SDS agents communicate hypotheses via a one-to-one communication strategy analogous to the tandem running procedure observed in some species of ant. A positive feedback mechanism ensures that, over time, a population of agents stabilise around the global-best solution. SDS is both an efficient and robust search and optimisation algorithm, which has been extensively mathematically described.

Applications

Swarm Intelligence-based techniques can be used in a number of applications. The U.S. military is investigating swarm techniques for controlling unmanned vehicles. The European Space Agency is thinking about an orbital swarm for self assembly and interferometry. NASA is investigating the use of swarm technology for planetary mapping. A 1992 paper by M. Anthony Lewis and George A. Bekeydiscusses the possibility of using swarm intelligence to control nanobots within the body for the purpose of killing cancer tumors. Swarm intelligence has also been applied for data mining.

Crowd Simulation

Artists are using swarm technology as a means of creating complex interactive systems or simulating crowds.

Stanley and Stella in: Breaking the Ice was the first movie to make use of swarm technology for rendering, realistically depicting the movements of groups of fish and birds using the Boids system. Tim Burton's *Batman Returns* also made use of swarm technology for showing the movements of a group of bats. The *Lord of the Rings* film trilogy made use of similar technology, known as Massive, during battle scenes. Swarm technology is particularly attractive because it is cheap, robust, and simple.

Airlines have used swarm theory to simulate passengers boarding a plane. Southwest Airlines researcher Douglas A. Lawson used an ant-based computer simulation employing only six interaction rules to evaluate boarding times using various boarding methods.(Miller, 2010, xii-xviii).

Ant-based Routing

The use of Swarm Intelligence in Telecommunication Networks has also been researched, in the form of Ant Based Routing. This was pioneered separately by Dorigo et al. and Hewlett Packard in the mid-1990s, with a number of variations since. Basically this uses a probabilistic routing table rewarding/reinforcing the route successfully traversed by each "ant" (a small control packet) which flood the network. Reinforcement of the route in the forwards, reverse direction and both simultaneously have been researched: backwards reinforcement requires a symmetric network and couples the two directions together; forwards reinforcement rewards a route before the outcome is known (but then you pay for the cinema before you know how good the film is). As the system behaves stochastically and is therefore lacking repeatability, there are large hurdles to commercial deployment. Mobile media and new technologies have the potential to change the threshold for collective action due to swarm intelligence (Rheingold: 2002, P175).

Airlines have also used ant-based routing in assigning aircraft arrivals to airport gates. At Southwest Airlines a software program uses swarm theory, or swarm intelligence—the idea that a colony of ants works better than one alone. Each pilot acts like an ant searching for the best airport gate. "The pilot learns from his experience what's the best for him, and it turns out that that's the best solution for the airline," Douglas A. Lawson explains. As a result, the "colony" of pilots always go to gates they can arrive at and depart from quickly. The program can even alert a pilot of plane back-ups before they happen. "We can anticipate that it's going to happen, so we'll have a gate available," Lawson says.

Gradient

In vector calculus, the gradient of a scalar field is a vector field that points in the direction of the greatest rate of increase of the scalar field, and whose magnitude is the greatest rate of change. A generalization of the gradient for functions on a Euclidean space that have values in another Euclidean space is the Jacobian. A further generalization for a function from one Banach space to another is the Fréchet derivative.

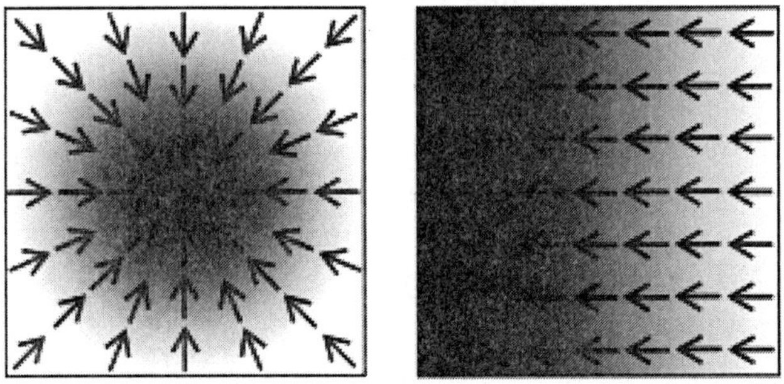

Figure: In the above two images, the scalar field is in black and white, black representing higher values, and its corresponding gradient is represented by blue arrows.

Interpretations

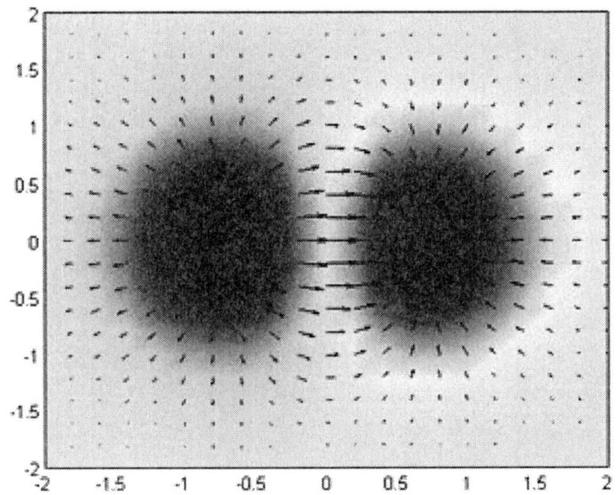

Figure: Gradient of the 2-d function $f(x, y) = xe^{-x^2-y^2}$ is plotted as blue arrows over the pseudocolor plot of the function

Consider a room in which the temperature is given by a scalar field, *T*, so at each point (*x*,*y*,*z*) the temperature is *T*(*x*,*y*,*z*). (We will assume that the temperature does not change over time.) At each point in the room, the gradient of *T* at that point will show the direction the temperature rises most quickly. The magnitude of the gradient will determine how fast the temperature rises in that direction. Consider a surface whose height above sea level at a point (x,y) is$H(x,y)$. The gradient of H at a point is a vector pointing in the direction of the steepest slope or grade at that point. The steepness of the slope at that point is given by the magnitude of the gradient vector.

The gradient can also be used to measure how a scalar field changes in other directions, rather than just the direction of greatest change, by taking a dot product. Suppose that the steepest slope on a hill is 40%. If a road goes directly up the hill, then the steepest slope on the road will also be 40%. If, instead, the road goes around the hill at an angle, then it will have a shallower slope. For example, if the angle between the road and the uphill direction, projected onto the horizontal plane, is 60°, then the steepest slope along the road will be 20%, which is 40% times the cosine of 60°.

This observation can be mathematically stated as follows. If the hill height function H isdifferentiable, then the gradient of Hdotted with a unit vector gives the slope of the hill in the direction of the vector. More precisely, when H is differentiable, the dot product of the gradient of H with a given unit vector is equal to the directional derivativeof H in the direction of that unit vector.

Definition

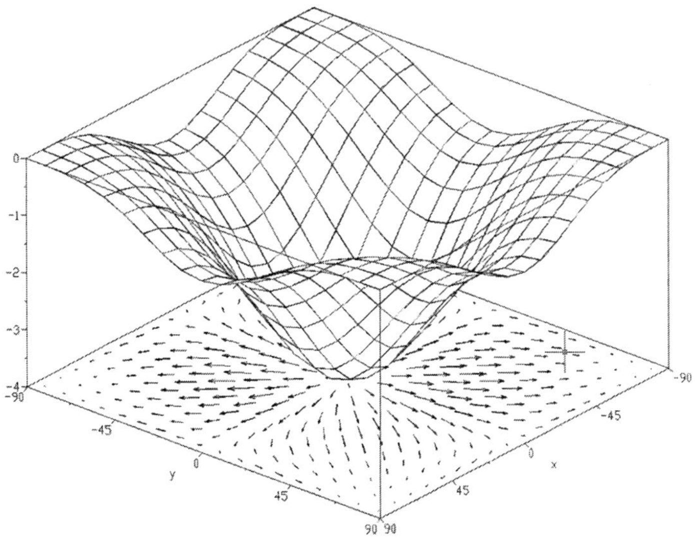

Figure: The gradient of the function $f(x,y) = -(\cos^2 x + \cos^2 y)^2$ depicted as a projected vector field on the bottom plane

The gradient (or gradient vector field) of a scalar function $f(x_1,x_2,x_3,\ldots,x_n)$ is denoted ∇f or $\vec{\nabla} f$ where ∇ (thenabla symbol) denotes the vectordifferential operator, del. The notation grad(f) is also used for the gradient. The gradient of f is defined to be the vector field whose components are the partial derivatives of f. That is:

$$\nabla f = \left(\frac{\partial f}{\partial x_1}, \ldots, \frac{\partial f}{\partial x_n} \right).$$

Here the gradient is written as a row vector, but it is often taken to be a column vector. When a function also depends on a parameter such as time, the gradient often refers simply to the vector of its spatial derivatives only.

The gradient of a vector $\mathbf{f} = (f_1, f_2, f_3)$ is $\nabla \mathbf{f} = \frac{\partial f_j}{\partial x_i} \mathbf{e}_i \mathbf{e}_j$

or the Jacobian matrix $\frac{\partial(f_1, f_2, f_3)}{\partial(x_1, x_2, x_3)}$

More generally, the gradient may be defined using the exterior derivative:

$$\nabla \mathbf{f} = (\mathbf{df})^{\sharp}$$

Here b and # are the musical isomorphisms.

Expression in 3-dimensional Rectangular Coordinates

The form of the gradient depends on the coordinate system used. In Cartesian coordinates, the above expression expands to

$$\nabla f(x,y,z) = \left(\frac{\partial f}{\partial x}, \frac{\partial f}{\partial y}, \frac{\partial f}{\partial z} \right)$$

which is often written using the standard unit vectors $\hat{\mathbf{i}}, \hat{\mathbf{j}}, \hat{\mathbf{k}}$ as

$$\frac{\partial f}{\partial x} \hat{\mathbf{i}} + \frac{\partial f}{\partial y} \hat{\mathbf{j}} + \frac{\partial f}{\partial z} \hat{\mathbf{k}}$$

Example

For example, the gradient of the function in Cartesian coordinates

$$f(x,y,z) = 2x + 3y^2 - \sin(z)$$

is:

$$\nabla f = \left(\frac{\partial f}{\partial x}, \frac{\partial f}{\partial y}, \frac{\partial f}{\partial z} \right) = \left(2, 6y, -\cos(z)\right).$$

Expression in 3-dimensional Spherical Polar Coordinates

$$\frac{\partial f}{\partial r}\hat{\mathbf{r}}+\frac{1}{r}\frac{\partial f}{\partial \theta}\hat{\theta}+\frac{1}{r\sin\theta}\frac{\partial f}{\partial \phi}\hat{\phi}$$

See Del in cylindrical and spherical coordinates.

Gradient and the Derivative or Differential

Linear Approximation to a Function

The gradient of a function f from the Euclidean space $\mathbb{R}^n$ to $\mathbb{R}$ at any particular point $x_0 \, \mathbb{R}^n$ in characterizes the best linear approximation to f at x_0. The approximation is as follows: $f(x) \approx f(x_0) + (\nabla f)_{x_0} \cdot (x - x_0)$ for x close to x_0, where $(\nabla f)_{x_0}$ is the gradient of $f$computed at x_0, and the dot denotes the dot product on $\mathbb{R}^n$. This equation is equivalent to the first two terms in the multi-variable Taylor Series expansion of f at x_0.

Differential or (Exterior) Derivative

The best linear approximation to a function $f : \mathbb{R}^n \to \mathbb{R}$ at a point x in $\mathbb{R}^n$ is a linear map from $\mathbb{R}^n$ to $\mathbb{R}$ which is often denoted by df_x or $Df(x)$ and called the differential or (total) derivative of $f$at x. The gradient is therefore related to the differential by the formula $(\nabla f)_x \cdot v = \mathrm{d}f_x(v)$ for any $v \in \mathbb{R}^n$. The function df, which maps x to df_x, is called the differential or exterior derivative of $f$and is an example of a differential 1-form.

If $\mathbb{R}^n$ is viewed as the space of (length n) column vectors (of real numbers), then one can regard df as the row vector

$$\mathrm{d}f = \left(\frac{\partial f}{\partial x_1}, \ldots, \frac{\partial f}{\partial x_n}\right)$$

so that d$f_x(v)$ is given by matrix multiplication. The gradient is then the corresponding column vector, i.e., $\nabla f = \mathrm{d}f^T$.

Gradient as a Derivative

Let U be an open set in R^n. If the function f: $U \to \mathrm{R}$ is differentiable, then the differential of f is the(Fréchet) derivative of f. Thus ∇f is a function from U to the space R such that

$$\lim_{h\to 0}\frac{\|f(x+h)-f(x)-\nabla f(x)\cdot h\|}{\|h\|}=0$$

where • is the dot product.

As a consequence, the usual properties of the derivative hold for the gradient:

Linearity

The gradient is linear in the sense that if f and g are two real-valued functions differentiable at the point $a \in \mathbf{R}^n$, and α and β are two constants, then $\alpha f + \beta g$ is differentiable at a, and moreover

$$\nabla(\alpha f + \beta g)(a) = \alpha \nabla f(a) + \beta \nabla g(a).$$

Product Rule

If f and g are real-valued functions differentiable at a point a"R^n, then the product rule asserts that the product $(fg)(x) = f(x)g(x)$ of the functions f and g is differentiable at a, and

$$\nabla(fg)(a) = f(a)\nabla g(a) + g(a)\nabla f(a).$$

Chain Rule

Suppose that $f{:}A \rightarrow R$ is a real-valued function defined on a subset A of R^n, and that f is differentiable at a point a. There are two forms of the chain rule applying to the gradient. First, suppose that the function g is a parametric curve; that is, a function $g : I \rightarrow R^n$ maps a subset $I \subset \mathbf{R}$ into R^n. If g is differentiable at a point $c \in I$ such that $g(c) = a$, then

$$(f^\circ g)'(c) = \nabla f(a) \cdot g'(c),$$

where $\circ$ is the composition operator. More generally, if instead I,"R^k, then the following holds:

$$\nabla(f^\circ g)(c) = (\nabla f(a))(Dg(c))^T$$

where $(Dg)^T$ denotes the transpose Jacobian matrix.

For the second form of the chain rule, suppose that $h : I \rightarrow \mathbf{R}$ is a real valued function on a subset $\boldsymbol{I}$ of $\mathbf{R}$, and that h is differentiable at the point $c = f(a) \in I$. Then

$$\nabla(h^\circ f)(a) = h'(c)\nabla f(a).$$

Transformation Properties

Although the gradient is defined in term of coordinates, it is contravariant under the application of anorthogonal matrix to the coordinates. This is true in the sense that if A is an orthogonal matrix, then

$$\nabla(f(Ax)) = A^T(\nabla f)(Ax) = A^{-1}(\nabla f)(Ax)$$

which follows by the chain rule above. A vector transforming in this way is known as a contravariant vector, and so the gradient is a special type of tensor.

The differential is more natural than the gradient because it is invariant under all coordinate transformations (or diffeomorphisms), whereas the gradient is only invariant under orthogonal transformations (because of the implicit use of the dot product in its definition). Because of this, it is common to blur the distinction between the two concepts using the notion of covariant and contravariant vectors. From this point of view, the components of the gradient transform covariantly under changes of coordinates, so it is called a covariant vector field, whereas the components of a vector field in the usual sense transform contravariantly. In this language the gradient *is* the differential, as a covariant vector field is the same thing as a differential 1-form.

Further Properties and Applications

Level sets: f the partial derivatives of f are continuous, then the dot product $(\nabla f)_x \cdot v$ of the gradient at a point$x$ with a vector v gives the directional derivative of f at x in the direction v. It follows that in this case the gradient of f is orthogonal to the level sets of f. For example, a level surface in three-dimensional space is defined by an equation of the form $F(x, y, z) = c$. The gradient of F is then normal to the surface.

More generally, any embedded hypersurface in a Riemannian manifold can be cut out by an equation of the form $F(P) = 0$ such that dF is nowhere zero. The gradient of F is then normal to the hypersurface.

Let us consider a function f at a point P. If we draw a surface through this point P and the function has the same value at all points on this surface,then this surface is called a 'level surface'.

Conservative Vector Fields

The gradient of a function is called a gradient field. A (continuous) gradient field is always aconservative vector field: its line integral along any path depends only on the endpoints of the path, and can be evaluated by the gradient theorem (the fundamental theorem of calculus for line integrals). Conversely, a (continuous) conservative vector field is always the gradient of a function.

Riemannian Manifolds

For any smooth function f on a Riemannian manifold (*M*,*g*), the gradient of *f* is the vector field ∇f such that for any vector field *X*,

$$g(\nabla f, X) = \partial_X f, \qquad \text{i.e.,} \quad g_x((\nabla f)_x, X_x) = (\partial_X f)(x)$$

where $g_x(\cdot,\cdot)$ denotes the inner product of tangent vectors at *x* defined by the metric *g* and $\partial_X f$ (sometimes denoted *X*(*f*)) is the function that takes any point *x*"*M* to the directional derivative of *f* in the direction *X*, evaluated at *x*. In other words, in a coordinate chart φ from an open subset of *M* to an open subset of R^n $(\partial_X f)(x)$ is given by:

$$\sum_{j=1}^{n} X^j(\varphi(x)) \frac{\partial}{\partial x_j}(f \circ \varphi^{-1})\Big|_{\varphi(x)},$$

where X^j denotes the *j*th component of *X* in this coordinate chart.

So, the local form of the gradient takes the form:

$$\nabla f = g^{ik} \frac{\partial f}{\partial x^k} \frac{\partial}{\partial x^i}.$$

Generalizing the case *M*=R^n, the gradient of a function is related to its exterior derivative, since $(\partial_X f)(x) = df_x(X_x)$ More precisely, the gradient ∇f is the vector field associated to the differential 1-form d*f* using the musical isomorphism $\sharp = \sharp^g : T^*M \to TM$ (called "sharp") defined by the metric *g*.

The relation between the exterior derivative and the gradient of a function onR^n is a special case of this in which the metric is the flat metric given by the dot product.

Non-cartesian Coordinate Systems

In cylindrical coordinates, the gradient is given by (Schey 1992, pp. 139–142):

$$\nabla f(\rho,\phi,z) = \frac{\partial f}{\partial \rho}\mathbf{e}_\rho + \frac{1}{\rho}\frac{\partial f}{\partial \phi}\mathbf{e}_\phi + \frac{\partial f}{\partial z}\mathbf{e}_z$$

where ϕ is2€he azimuthal angle, *z* is the axial coordinate, and e_ρ, e_φ and e_z are unit vectors pointing along the coordinate directions.

In spherical coordinates (Schey 1992, pp. 139–142):

$$\nabla f(r,\theta,\phi) = \frac{\partial f}{\partial r}\mathbf{e}_r + \frac{1}{r}\frac{\partial f}{\partial \theta}\mathbf{e}_\theta + \frac{1}{r\sin\theta}\frac{\partial f}{\partial \phi}\mathbf{e}_\phi$$

where ϕ is the azimuth angle and θ is the zenith angle.

Gradient Descent

Gradient descent is a first-order optimization algorithm. To find a local minimum of a function using gradient descent, one takes steps proportional to the *negative* of the gradient (or of the approximate gradient) of the function at the current point. If instead one takes steps proportional to the *positive* of the gradient, one approaches a local maximum of that function; the procedure is then known asgradient ascent.

Gradient descent is also known as steepest descent, or the method of steepest descent. When known as the latter, gradient descent should not be confused with the method of steepest descent for approximating integrals.

Description

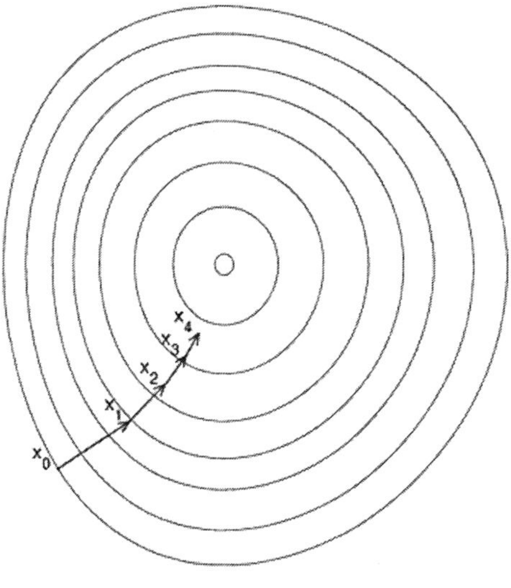

Figure: Illustration of gradient descent

Gradient descent is based on the observation that if the multivariable function $F(\mathbf{x})$ is defined anddifferentiable in a neighborhood of a point a, then $F(\mathbf{x})$ decreases*fastest* if one goes from in the direction of the negative gradient of Fat a$-\nabla F(\mathbf{a})$. It follows that, if

$$\mathbf{b} = \mathbf{a} - \gamma \nabla F(\mathbf{a})$$

for $\gamma > 0$ a small enough number, then $F(\mathbf{a}) \geq F(\mathbf{b})$ With this observation in mind, one starts with a guess $\mathbf{x}_0$ for a local minimum of$F$, and considers the sequence $\mathbf{x}_0, \mathbf{x}_1, \mathbf{x}_2, \ldots$ such that

$$\mathbf{x}_{n+1} = \mathbf{x}_n - \gamma_n \nabla F(\mathbf{x}_n), n \geq 0.$$

We have

$$F(\mathbf{x}_0) \geq F(\mathbf{x}_1) \geq F(\mathbf{x}_2) \geq \cdots,$$

so hopefully the sequence $(\mathbf{x}_n)$ converges to the desired local minimum. Note that the value of the*step size* γ is allowed to change at every iteration. This process is illustrated in the picture to the right. Here F is assumed to be defined on the plane, and that its graph has a bowl shape. The blue curves are the contour lines, that is, the regions on which the value of F is constant. A red arrow originating at a point shows the direction of the negative gradient at that point. Note that the (negative) gradient at a point is orthogonal to the contour line going through that point. We see that gradient *descent* leads us to the bottom of the bowl, that is, to the point where the value of the function F is minimal.

Examples

Gradient descent has problems with pathological functions such as the Rosenbrock function shown here.

$$f(x_1,x_2) = (1-x_1)^2 + 100(x_2 - x_1^2)^2.$$

The Rosenbrock function has a narrow curved valley which contains the minimum. The bottom of the valley is very flat. Because of the curved flat valley the optimization is zig-zagging slowly with small stepsizes towards the minimum.

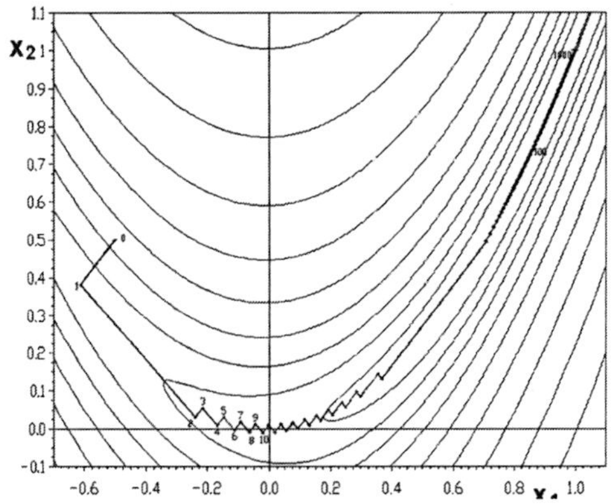

The "Zig-Zagging" nature of the method is also evident below, where the gradient ascent method is applied to

$$F(x,y) = \sin\left(\frac{1}{2}x^2 - \frac{1}{4}y^2 + 3\right)\cos(2x + 1 - e^y)$$

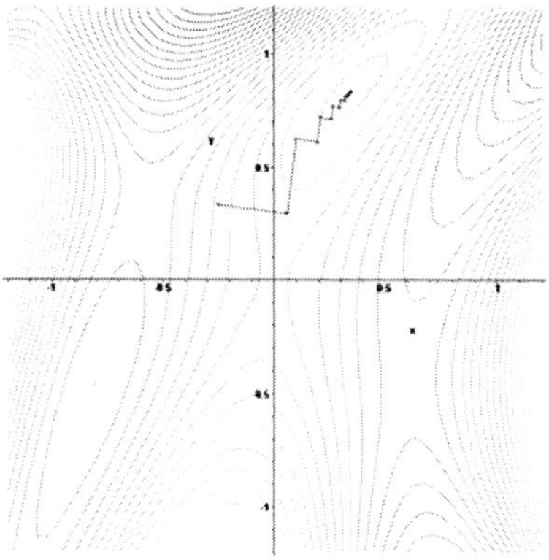

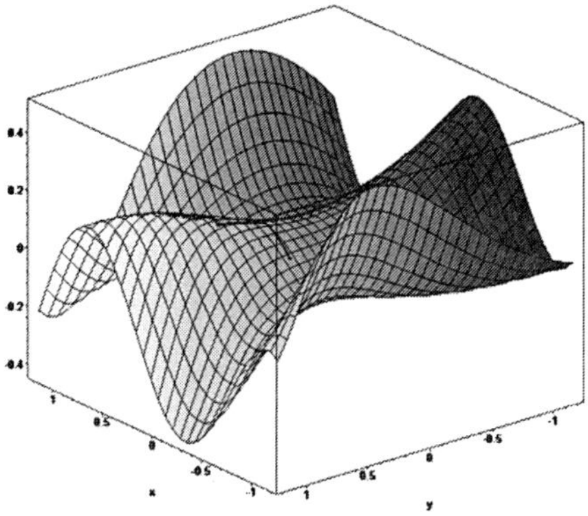

Limitations

For some of the above examples, gradient descent is relatively slow close to the minimum: Technically, its asymptotic rate of convergence is inferior to other methods. For poorly conditioned convex problems, gradient descent increasingly 'zigzags' as the gradients point nearly orthogonally to the shortest direction to a minimum point. For more details, see Comments below:

For non-differentiable functions, gradient methods are ill-defined. For locally Lipschitz problems and especially for convex minimization problems, bundle methods of descent are well-defined. Non-descent methods, like subgradient projection methods, may also be used.

Solution of a Linear System

Gradient descent can be used to solve a system of linear equations, reformulated as a quadratic minimization problem, e.g., using linear least squares. Solution of

$$A\mathbf{x} - \mathbf{b} = 0$$

in the sense of linear least squares is defined as minimizing the function

$$F(x) = \left\|A\mathbf{x} - \mathbf{b}^2.\right\|$$

In traditional linear least squares for real A and b the Euclidean norm is used, in which case

$$\nabla F(\mathbf{x}) = 2A^T(A\mathbf{x} - \mathbf{b}).$$

In the case that A is real, square, symmetric and positive definite, a different popular choice of the norm is

$$\left\|\mathbf{a}^2 = \mathbf{a}^T A^{-1}\mathbf{a}\right\|$$

which produces a different equation with a better condition number:

$$\nabla F(\mathbf{x}) = 2(A\mathbf{x} - \mathbf{b}).$$

In either case, the line search minimization, finding the locally optimal step size γ on every iteration, can be performed analytically, and explicit formulas for the locally optimal γ are known.

Solution of a non-linear System

Gradient descent can also be used to solve a system of nonlinear equations. Below is an example that shows how to use the gradient descent to solve for three unknown variables, x_1, x_2, and x_3. This example shows one iteration of the gradient descent.

Consider a nonlinear system of equations:

$$\begin{cases} 3x_1 - \cos(x_2x_3) - \tfrac{3}{2} = 0 \\ 4x_1^2 - 625x_2^2 + 2x_2 - 1 = 0 \\ \exp(-x_1x_2) + 20x_3 + \tfrac{10\pi-3}{3} = 0 \end{cases}$$

suppose we have the function

$$G(\mathbf{x}) = \begin{bmatrix} 3x_1 - \cos(x_2x_3) - \tfrac{3}{2} \\ 4x_1^2 - 625x_2^2 + 2x_2 - 1 \\ \exp(-x_1x_2) + 20x_3 + \tfrac{10\pi-3}{3} \end{bmatrix}$$

where

$$\mathbf{x} = \begin{bmatrix} x_1 \\ x_2 \\ x_3 \end{bmatrix}$$

and the objective function

$$F(\mathbf{x}) = \tfrac{1}{2} G^{\mathrm{T}}(\mathbf{x}) G(\mathbf{x})$$

$$= \tfrac{1}{2}((3x_1 - \cos(x_2x_3) - \tfrac{3}{2})^2 + (4x_1^2 - 625x_2^2 + 2x_2 - 1)^2 + (\exp(-x_1x_2) + 20x_3 + \tfrac{10\pi-3}{3})^2)$$

With initial guess

$$\mathbf{x}^{(0)} = \begin{bmatrix} x_1 \\ x_2 \\ x_3 \end{bmatrix} = \begin{bmatrix} 0 \\ 0 \\ 0 \end{bmatrix}$$

We know that

$$\mathbf{x}^{(1)} = \mathbf{x}^{(0)} - \gamma_0 \nabla F(x^{(0)})$$

where

$$\nabla F(\mathbf{x}^{(0)}) = J_G(\mathbf{x}^{(0)})^{\mathrm{T}} * G(\mathbf{x}^{(0)})$$

The Jacobian matrix $J_G(\mathbf{x}^{(0)})$

$$J_G = \begin{bmatrix} 3 & \sin(x_2x_3)x_3 & \sin(x_2x_3)x_2 \\ 8x_1 & -1250x_2 + 2 & 0 \\ -x_2\exp(-x_1x_2) & -x_1\exp(-x_1x_2) & 20 \end{bmatrix}$$

Then evaluating these terms at $\mathbf{x}^{(0)}$

$$J_G\left(\mathbf{x}^{(0)}\right) = \begin{bmatrix} 3 & 0 & 0 \\ 0 & 2 & 0 \\ 0 & 0 & 20 \end{bmatrix}$$

and

$$G(\mathbf{x}^{(0)}) = \begin{bmatrix} -2.5 \\ -1 \\ 10.472 \end{bmatrix}$$

So that

$$\mathbf{x}^{(1)} = 0 - \gamma_0 \begin{bmatrix} -7.5 \\ -2 \\ 209.44 \end{bmatrix}.$$

and

$$F\left(\mathbf{x}^{(0)}\right) = 0.5((-2.5)^2 + (-1)^2 + (10.472)^2) = 58.456$$

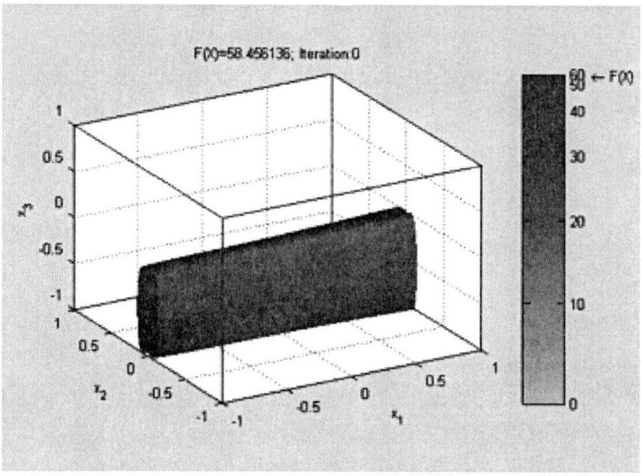

An animation showing the first 83 iterations of gradient decent applied to this example. Surfaces are isosurfaces of $F(\mathbf{x}^{(n)})$ at current guess $\mathbf{x}^{(n)}$ and arrows show the direction of descent. Due to a small and constant step size, the convergence is slow.

Now a suitable γ_0 must be found such that $F(\mathbf{x}^{(1)}) \le F(\mathbf{x}^{(0)})$. This can be done with any of a variety of line search algorithms. One might also simply guess $\gamma_0 = 0.001$which gives

$$\mathbf{x}^{(1)} = \begin{bmatrix} 0.0075 \\ 0.002 \\ -0.20944 \end{bmatrix}$$

evaluating at this value,

$$F\left(\mathbf{x}^{(1)}\right) = 0.5((-2.48)^2 + (-1.00)^2 + (6.28)^2) = 23.306$$

The decrease from $F(\mathbf{x}^{(0)}) = 58.456$ to the next step's value of $F(\mathbf{x}^{(1)}) = 23.306$ is a sizable decrease in the objective function. Further steps would reduce its value until a solution to the system was found.

Comments

Gradient descent works in spaces of any number of dimensions, even in infinite-dimensional ones. In the latter case the search space is typically a function space, and one calculates the Gβteaux derivative of the functional to be minimized to determine the descent direction.

The gradient descent can take many iterations to compute a local minimum with a required accuracy, if the curvature in different directions is very different for the given function.

For such functions,preconditioning, which changes the geometry of the space to shape the function level sets likeconcentric circles, cures the slow convergence. Constructing and applying preconditioning can be computationally expensive, however.

The gradient descent can be combined with a line search, finding the locally optimal step size γ on every iteration. Performing the line search can be time-consuming. Conversely, using a fixed small γcan yield poor convergence.

Methods based on Newton's method and inversion of the Hessian using conjugate gradient techniques can be better alternatives. Generally, such methods converge in fewer iterations, but the cost of each iteration is higher. An example is the BFGS method which consists in calculating on every step a matrix by which the gradient vector is multiplied to go into a "better" direction, combined with a more sophisticated line search algorithm, to find the "best" value of γ. For extremely large problems, where the computer memory issues dominate, a limited-memory method such as L-BFGS should be used instead of BFGS or the steepest descent.

Gradient descent can be viewed as Euler's method for solving ordinary differential equations $x'(t) = -\nabla f(x(t))$ of a gradient flow.

A Computational Example

The gradient descent algorithm is applied to find a local minimum of the function $f(x)=x^4-3x^3+2$, with derivative $f'(x)=4x^3-9x^2$. Here is an implementation in the Python scripting language.

```
# From calculation, we expect that the local minimum occurs at x=9/4
x_old = 0
x_new = 6 # The algorithm starts at x=6
eps = 0.01 # step size
precision = 0.00001
def f_prime(x):
return 4 * x**3 - 9 * x**2
while abs(x_new - x_old) > precision:
x_old = x_new
 x_new = x_old - eps * f_prime(x_new)
print "Local minimum occurs at ", x_new
```

The above piece of code has to be modified with regard to step size according to the system at hand and convergence can be made faster by using an adaptive step size. In the above case the step size is not adaptive. It stays at 0.01 in all the directions which can sometimes cause the method to fail by diverging from the minimum.

Candidate Solution

In optimization (a branch of mathematics), acandidate solution is a member of a set of possible solutions to a given problem. A candidate solution does not have to be a likely or reasonable solution to the problem – it is simply in the set that satisfies allconstraints.

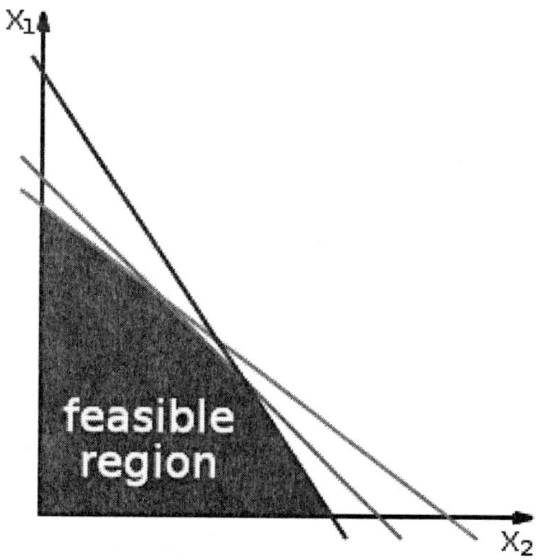

Figure: A series of linear programming constraints on two variables produces a region of possible values for those variables. Solvable problems will have a feasible region in the shape of a simple polygon.

The space of all candidate solutions is called thefeasible region, feasible set, search space, orsolution space.

In the case of the genetic algorithm, the candidate solutions are the individuals in the population being evolved by the algorithm.

Meta-optimization

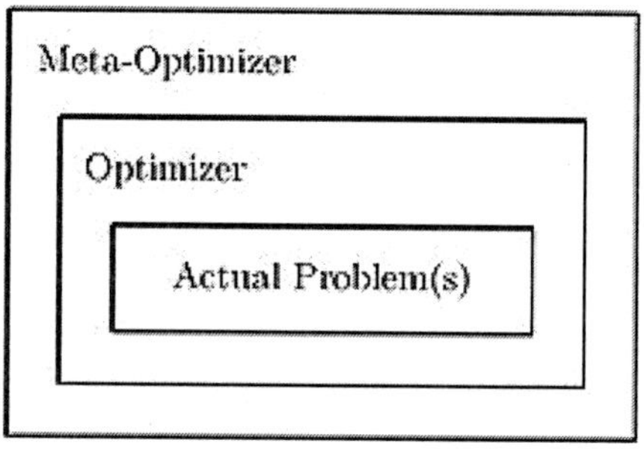

Figure: Meta-optimization concept.

In numerical optimization, meta-optimization is the use of one optimization method to tune another optimization method. Meta-optimization is reported to have been used as early as in the late 1970s by Mercer and Sampson for finding optimal parameter settings of a genetic algorithm. Meta-optimization is also known in the literature as meta-evolution, super-optimization, automated parameter calibration, hyper-heuristics, etc.

Motivation

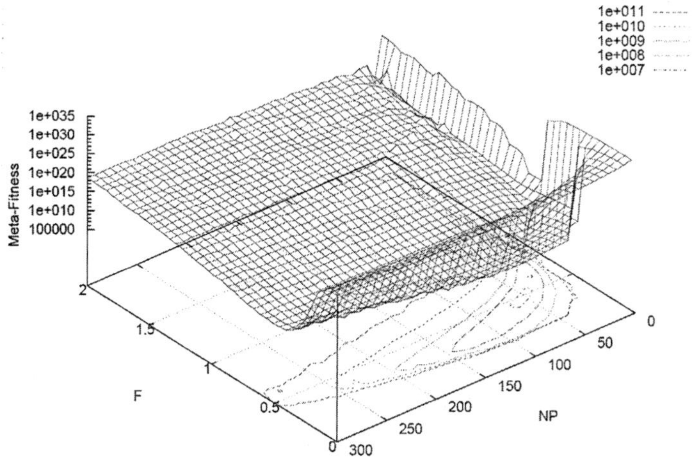

Figure: Performance landscape for differential evolution.

Optimization methods such as genetic algorithm anddifferential evolution have several parameters that govern their behaviour and efficacy in optimizing a given problem and these parameters must be chosen by the practitioner to achieve satisfactory results. Selecting the behavioural parameters by hand is a laborious task that is susceptible to human misconceptions of what makes the optimizer perform well.

The behavioural parameters of an optimizer can be varied and the optimization performance plotted as a landscape. This is computationally feasible for optimizers with few behavioural parameters and optimization problems that are fast to compute, but when the number of behavioural parameters increases the time usage for computing such a performance landscape increases exponentially. This is the curse of dimensionality for the search-space consisting of an optimizer's behavioural parameters. An efficient method is therefore needed to search the space of behavioural parameters.

Methods

A simple way of finding good behavioural parameters for an optimizer is to employ another overlaying optimizer, called the meta-optimizer. There are different ways of doing this depending on whether the behavioural parameters to be tuned are real-valued or discrete-valued, and depending on what performance measure is being used, etc.

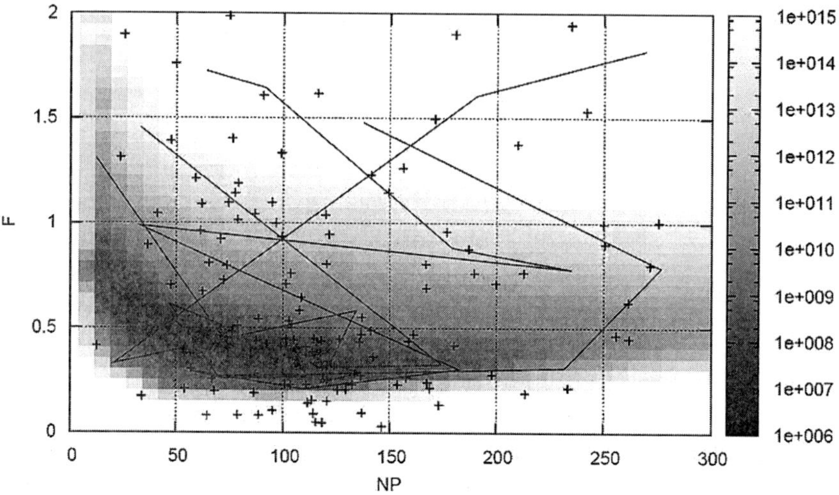

Figure: Meta-optimization of differential evolution.

Meta-optimizing the parameters of a genetic algorithm was done by Grefenstette and Keane, amongst others, and experiments with meta-optimizing both the parameters and the genetic operators were reported by Bδck. Meta-optimization of particle swarm optimization was done by Meissner et al. as well as by Pedersen and Chipperfield, who also meta-optimized differential evolution. Birattari et al. meta-optimized ant colony optimization. Statistical models have also been used to reveal more about the relationship between choices of behavioural parameters and optimization performance, see for example Francois and Lavergne, and Nannen and Eiben. A comparison of various meta-optimization techniques was done by Smit and Eiben.

Multi-objective Optimization

Multi-objective optimization (or multi-objective programming), also known as multi-criteriaor multi-attribute optimization, is the process of simultaneously optimizing two or more conflicting objectives subject to certain constraints.

Multiobjective optimization problems can be found in various fields: product and process design, finance, aircraft design, the oil and gas industry, automobile design, or wherever optimal decisions need to be taken in the presence of trade-offs between two or more conflicting objectives. Maximizing profit and minimizing the cost of a product; maximizing performance and minimizing fuel consumption of a vehicle; and minimizing weight while maximizing the strength of a particular component are examples of multi-objective optimization problems. For nontrivial multiobjective problems, one cannot identify a single solution that simultaneously optimizes each objective. While searching for solutions, one reaches points such that, when attempting to improve an objective further, other objectives suffer as a result. A tentative solution is called non-dominated, Pareto optimal, or Pareto efficient if it cannot be eliminated from consideration by replacing it with another solution which improves an objective without worsening another one. Finding such non-dominated solutions, and quantifying the trade-offs in satifying the different objectives, is the goal when setting up and solving a multiobjective optimization problem.

In mathematical terms, the multiobjective problem can be written as:

$$\begin{aligned} &\min_x \quad [\mu_1(x), \mu_2(x), \ldots, \mu_n(x)]^T \\ &\text{s.t.} \\ &g(x) \quad \leq 0 \\ &h(x) \quad = 0 \\ &x_l \leq \quad x \leq x_u \end{aligned}$$

where μ_i is the i-th objective function, g and h are the inequality and equality constraints, respectively, and x is the vector of optimization or decision variables. The solution to the above problem is a set of Pareto points. Thus, instead of being a unique solution to the problem, the solution to a multiobjective problem is a possibly infinite set of Pareto points.

A design point in objective space μ is termed Pareto optimal if there does not exist another feasible design objective vector μ^* such that $\mu_i \leq \mu_i^*$ for all $i \in \{1,2,\ldots,n\}$, and $\mu_j < \mu_j^*$ for at least one index of $j,\ j \in \{1,2,\ldots,n\}$

Solution Methods

There exist many methods to finding a solution to a multiobjective optimization problem, some of which are explained below:

Constructing a Single Aggregate Objective Function (AOF)

This is an intuitive approach to solving the multi-objective problem. The basic idea is to combine all of the objectives into a single objective function, called the AOF, such as the well-known weighted linear sum of the objectives. This objective function is optimized subject to technological constraints specifying how much of one objective must be sacrificed, from any given starting point, in order to gain a certain amount regarding the other objective. These technological constraints frequently come in the form $f(x_1,x_2) \geq 0$ for some function *f*, where x_1 and x_2 are the objectives (e.g., strength and lightness of a product).

Often the aggregate objective function is not linear in the objectives, but rather is non-linear, expressing increasing marginal dissatisfaction with greater incremental sacrifices in the value of either objective. Furthermore, sometimes the aggregate objective function is additively separable, so that it is expressed as a weighted average of a non-linear function of one objective and a non-linear function of another objective.

Then the optimal solution obtained will depend on the relative values of the weights specified. For example, if one is trying to maximize the strength of a machine component and minimize the production cost, and if a higher weight is specified for the cost objective compared to the strength, the solution will be one that favors lower cost over higher strength.

The weighted sum method, like any method of selecting a single solution as preferable to all others, is essentially subjective, in that a decision manager needs to supply the weights. Moreover, this approach may prove difficult to implement if the Pareto frontier is not globally convex and/or the objective function to be minimized is not globally concave.

The objective way of characterizing multi-objective problems, by identifying multiple Pareto optimal candidate solutions, requires a Pareto-compliant ranking method, favoring non-dominated solutions, as seen in current multi-objective evolutionary approaches such as NSGA-II and SPEA2. Here, no weight is required and thus no *a priori* information on the decision-maker's preferences is needed.However, to decide upon one of the Pareto-efficient options as the one to adopt requires information about the decision-maker's preferences. Thus the objective characterization of the problem is simply the first stage in a two-stage analysis, consisting of (1) identifying the non-dominated possibilities, and (2) choosing among them.

The NBI, NC, SPO and DSD Methods

The Normal Boundary Intersection (NBI), Normal Constraint (NC), Successive Pareto Optimization (SPO), and Directed Search Domain (DSD) methods solve the multi-objective optimization problem by constructing several AOFs. The solution of each AOF yields a Pareto point, whether locally or globally.

The NC and DSD methods suggest two different filtering procedures to remove locally Pareto points. The AOFs are constructed with the target of obtaining evenly distributed Pareto points that give a good impression (approximation) of the real set of Pareto points.

The DSD, NC and SPO methods generate solutions that represent some peripheral regions of the set of Pareto points for more than two objectives that are known to be not represented by the solutions generated with the NBI method.

According to Erfani and Utyuzhnikov, the DSD method works reasonably more efficiently than its NC and NBI counterparts on some difficult test cases in the literature.

Evolutionary Algorithms

Evolutionary algorithms are popular approaches to solving multiobjective optimization. Currently most evolutionary optimizers apply Pareto-based ranking schemes. Genetic algorithms such as the Non-dominated Sorting Genetic Algorithm-II (NSGA-II) and Strength Pareto Evolutionary Algorithm 2 (SPEA-2) have become standard approaches, although some schemes based on particle swarm optimization and simulated annealing are significant. The main advantage of evolutionary algorithms, when applied to solve multi-objective optimization problems, is the fact that they typically optimize sets of solutions, allowing computation of an approximation of the entire Pareto front in a single algorithm run. The main disadvantage of evolutionary algorithms is the much lower speed.

Other Methods

- Multiobjective Optimization using Evolutionary Algorithms (MOEA).
- PGEN (Pareto surface generation for convex multiobjective instances)
- IOSO (Indirect Optimization on the basis of Self-Organization).

Economics

In economics, the study of resource allocation under scarcity, many problems involve multiple objectives along with constraints on what combinations of those objectives are attainable. For example, a consumer's demands for various goods are determined by the process of maximization of the utility derived from those goods, subject to a constraint based on how much income is available to spend on those goods and on the prices of those goods. This constraint allows more of one good to be purchased only at the sacrifice of consuming less of another good; therefore, the various objectives (more consumption of each good is preferred) are in conflict with each other according to this constraint. A common method for analyzing such a problem is to use a graph of indifference curves, representing preferences, and a budget constraint, representing the trade-offs that the consumer is faced with.

Another example involves the production possibilities frontier, which specifies what combinations of various types of goods can be produced by a society with certain amounts of various resources. The frontier specifies the trade-offs that the society is faced with — if the society is fully utilizing its resources, more of one good can be produced only at the expense of producing less of another good. A society must then use some process to choose among the possibilities on the frontier.

Macroeconomic policy-making is a context requiring multi-objective optimization. Typically a central bank must choose a stance for monetary policy that balances competing objectives — low inflation, low unemployment, low balance of trade deficit, etc. To do this, the central bank uses a model of the economy that quantitatively describes the various causal linkages in the economy; it simulates the model repeatedly under various possible stances of monetary policy, in order to obtain a menu of possible predicted outcomes for the various variables of interest. Then in principle it can use an aggregate objective function to rate the alternative sets of predicted outcomes, although in practice central banks use a non-quantitative, judgement-based, process for ranking the alternatives and making the policy choice.

Finance

In finance, a common problem is to choose a portfolio when there are two conflicting objectives — the desire to have the expected value of portfolio returns be as high as possible, and the desire to haverisk, measured by the standard deviation of portfolio returns, be as low as possible. This problem is often represented by a graph in which the efficient frontier shows the best combinations of risk and expected return that are available, and in which indifference curves show the investor's preferences for various risk-expected return combinations. The problem of optimizing a function of the expected value (first moment) and the standard deviation (square root of the second moment) of portfolio return is called a two-moment decision model.

Linear Programming Applications

In linear programming problems, a linear objective function is optimized subject to linear constraints. Typically multiple variables of concern appear in the objective function. A vast body of research has been devoted to methods of solving these problems. Because the efficient set, the set of combinations of values of the various variables of interest having the feature that none of the variables can be given a better value without hurting the value of another variable, is piecewise linear and not continuouslydifferentiable, the problem is not dealt with by first specifying all the points on the Pareto-efficient set; instead, solution procedures utilize the aggregate objective function right from the start.

Many practical problems in operations research can be expressed as linear programming problems. Certain special cases of linear programming, such as network flow problems and multi-commodity flow problems are considered important enough to have generated much research on specialized algorithms for their solution. Linear programming is heavily used in microeconomics and companymanagement, for dealing with such issues as planning, production, transportation, technology, and so forth.

Optimal Control Applications

In engineering and economics, many problems involve multiple objectives which are not describable as the-more-the-better or the-less-the-better; instead, there is an ideal target value for each objective, and the desire is to get as close as possible to the desired value of each objective. For example, one might want to adjust a rocket's fuel usage and orientation so that it arrives both at a specified place and at a specified time; or one might want to conduct open market operations so that both the inflation rate and the unemployment rate are as close as possible to their desired values.

Often such problems are subject to linear equality constraints that prevent all objectives from being simultaneously perfectly met, especially when the number of controllable variables is less than the number of objectives and when the presence of random shocks generates uncertainty. Commonly a multi-objective quadratic objective function is used, with the cost associated with an objective rising quadratically with the distance of the objective from its ideal value. Since these problems typically involve adjusting the controlled variables at various points in time and/or evaluating the objectives at various points in time, intertemporal optimization techniques are employed.

Pareto Efficiency

Pareto efficiency, or Pareto optimality, is a concept in economics with applications in engineeringand social sciences. The term is named after Vilfredo Pareto, an Italian economist who used the concept in his studies of economic efficiency and income distribution.

Given an initial allocation of goods among a set of individuals, a change to a different allocation that makes at least one individual better off without making any other individual worse off is called a Pareto improvement. An allocation is defined as "Pareto efficient" or "Pareto optimal" when no further Pareto improvements can be made.

Pareto efficiency is a minimal notion of efficiency and does not necessarily result in a socially desirable distribution of resources: it makes no statement about equality, or the overall well-being of a society.

Pareto Efficiency in Short

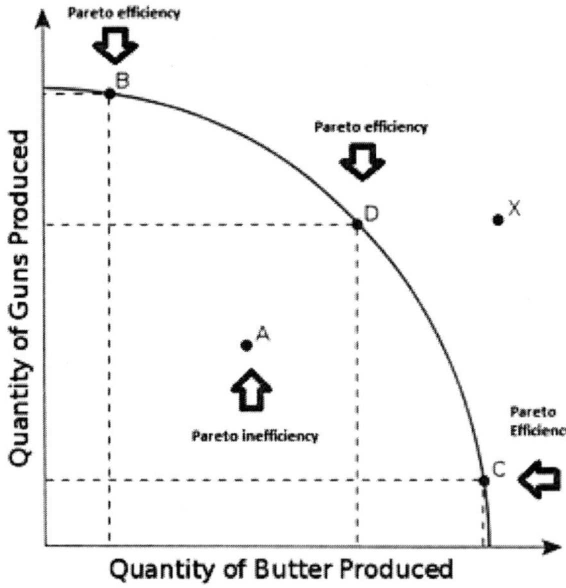

Looking at the Production-possibility frontier, shows how productive efficiency is a precondition for Pareto efficiency. Point A is not efficient in production because you can produce more of either one or both goods (Butter and Guns) without producing less of the other. Thus, moving from A to D enables you to make one person better off without making anyone else worse off (Pareto improvement). Moving to point C from point A, however, is not Pareto efficient, as fewer guns are produced. Likewise, moving to point B from point A is not Pareto efficient, as less butter is produced. A point on the frontier curve with the same x or y coordinate will be Pareto efficient.

An economic system that is *not* Pareto efficient implies that a certain change in allocation of goods (for example) may result in some individuals being made "better off" with no individual being made worse off, and therefore can be made more Pareto efficient through a Pareto improvement. Here 'better off' is often interpreted as "put in a preferred position." It is commonly accepted that outcomes that are *not* Pareto efficient are to be avoided, and therefore Pareto efficiency is an important criterion for evaluating economic systems and public policies.

If economic allocation in any system is not Pareto efficient, there is potential for a Pareto improvement—an increase in Pareto efficiency: through reallocation, improvements to at least one participant's well-being can be made better without reducing any other participant's well-being.

In the real world ensuring that nobody is disadvantaged by a change aimed at improving economic efficiency may require compensation of one or more parties. For instance, if a change in economic policy dictates that a legally protected monopoly ceases to exist and that market subsequently becomes competitive and more efficient, the monopolist will be made worse off. However, the loss to the monopolist will be more than offset by the gain in efficiency. This means the monopolist can be compensated for its loss while still leaving an efficiency gain to be realized by others in the economy. Thus, the requirement of nobody being made worse off for a gain to others is met. In real-world practice compensations have substantial frictional costs. They can also lead to incentive distortions over time since most real-world policy changes occur with players who are not atomistic, rather who have considerable market power (or political power) over time and may use it in agame theoretic manner. Compensation attempts may therefore lead to substantial practical problems of misrepresentation and moral hazard and considerable inefficiency as players behave opportunistically and with guile. In real-world practice, the compensation principle often appealed to is hypothetical. That is, for the alleged Pareto improvement (say from public regulation of the monopolist or removal of tariffs) some losers are not (fully) compensated. The change thus results in distribution effects in addition to any Pareto improvement that might have taken place. The theory of hypothetical compensation is part ofKaldor-Hicks efficiency, also called Potential Pareto Criterion.

Under certain idealized conditions, it can be shown that a system of free markets will lead to a Pareto efficient outcome. This is called the first welfare theorem. It was first demonstrated mathematically by economists Kenneth Arrow and Gerard Debreu. However, the result does not rigorously establish welfare results for real economies because of the restrictive assumptions necessary for the proof (markets exist for all possible goods, all markets are in full equilibrium, markets are perfectly competitive, transaction costs are negligible, there must be no externalities, and market participants must have perfect information). Moreover, it has since been demonstrated mathematically that, in the absence of perfect information or complete markets, outcomes will generically be Pareto inefficient (the Greenwald-Stiglitz theorem).

Pareto Improvements and Microeconomic Theory

Note that microeconomic analysis does not assume additive utility nor does it assume any interpersonal utility tradeoffs. To engage in interpersonal utility tradeoffs leads to greater good problems faced by earlier utilitarians. It also creates a question as to how weights are assigned and who assigns them, as well as questions regarding how to compare pleasure or pain across individuals.

Efficiency – in all of standard microeconomics – therefore refers to the absence of Pareto improvements. It does not in any way opine on the fairness of the allocation (in the sense ofdistributive justice or equity). An 'efficient' equilibrium could be one where one player has all the goods and other players have none (in an extreme example).

Weak and Strong Pareto Optimum

A "weak Pareto optimum" (WPO) nominally satisfies the same standard of not being Pareto-inferior to any other allocation, but for the purposes of weak Pareto optimization, an alternative allocation is considered to be a Pareto improvement *only if* the alternative allocation is strictly preferred by *all*individuals. In other words, when an allocation is WPO there are no possible alternative allocations whose realization would cause every individual to gain.

Weak Pareto-optimality is "weaker" than strong Pareto-optimality in the sense that the conditions for WPO status are "weaker" than those for SPO status: any allocation that can be considered an SPO will also qualify as a WPO, but a WPO allocation won't necessarily qualify as an SPO.

Under any form of Pareto-optimality, for an alternative allocation to be Pareto-superior to an allocation being tested—and, therefore, for the feasibility of an alternative allocation to serve as proof that the tested allocation is not an optimal one—the feasibility of the alternative allocation must show that the tested allocation fails to satisfy at least one of the requirements for SPO status. One may apply the same metaphor to describe the set of requirements for WPO status as being "weaker" than the set of requirements for SPO status. (Indeed, because the SPO set entirely encompasses the WPO set, with respect to *any* property the requirements for SPO status are of strength equal to or greater than the strength of the requirements for WPO status. Therefore, the requirements for WPO status are not merely weaker on balance or weaker according to the odds; rather, one may describe them more specifically and quite fittingly as "Pareto-weaker.")

- Note that when one considers the requirements for an *alternative* allocation's superiority according to one definition against the requirements for its superiority according to the other, the comparison between the requirements of the respective definitions is the *opposite* of the comparison between the requirements for optimality: To demonstrate the WPO-inferiority of an allocation being tested, an alternative allocation must falsify at least one of the particular conditions in the WPO subset, rather than merely falsify at least one of either these conditions or the other SPO conditions. Therefore, the requirements for weak Pareto-*superiority* of an *alternative* allocation are harder to satisfy (in other words, "stronger") than are the requirements for strong Pareto-superiority of an alternative allocation.
- It further follows that every SPO is a WPO (but not every WPO is an SPO): Whereas the WPO description applies to *any* allocation from which every feasible departure results in the

NON-IMPROVEMENT of at least one individual, the SPO description applies to *only* those allocations that meet *both* the WPO requirement *and* the more specific ("stronger") requirement that at least one non-improving individual exhibit a specific type of non-improvement, namely doing worse.

- The "strong" and "weak" descriptions of optimality continue to hold true when one construes the terms in the context set by the field of semantics: If one describes an allocation as being a WPO, one makes a "weaker" statement than one would make by describing it as an SPO: If the statements "Allocation X is a WPO" and "Allocation X is a SPO" are both true, then the former statement is less controversial than the latter in that to defend the latter, one must prove everything to defend the former "and then some." By the same token, however, the former statement is less informative or contentful in that it "says less" about the allocation; that is, the former statement contains, implies, and (when stated) asserts fewer constituent propositions about the allocation.

Formal Representation

Formally, a (strong/weak) Pareto optimum is a maximal element for the partial order relation of Pareto improvement/strict Pareto improvement: it is an allocation such that no other allocation is "better" in the sense of the order relation.

Pareto Frontier

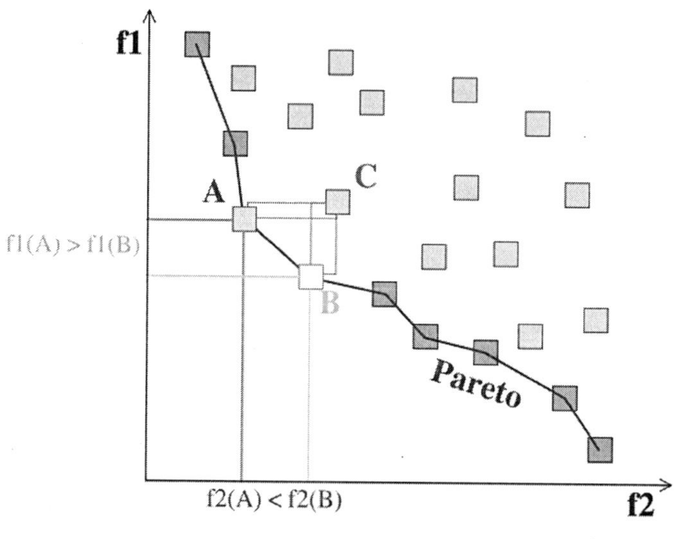

Example of a Pareto frontier. The boxed points represent feasible choices, and smaller values are preferred to larger ones. Point C is not on the Pareto Frontier because it is dominated by both point A and point B. Points A and B are not strictly dominated by any other, and hence do lie on the frontier.

Given a set of choices and a way of valuing them, the Pareto frontier or Pareto set orPareto front is the set of choices that are Pareto efficient. The Pareto frontier is particularly useful in engineering: by restricting attention to the set of choices that are Pareto-efficient, a designer can make tradeoffs within this set, rather than considering the full range of every parameter.

The Pareto frontier is defined formally as follows.

Consider a design space with n real parameters, and for each design space point there are m different criteria by which to judge that point. Let $f : \mathbb{R}^n \rightarrow \mathbb{R}^m$ be the function which assigns, to each design space point x, a criteria space point f(x). This represents the way of valuing the designs. Now, it may be that some designs are infeasible; so let X be a set of feasible designs in $\mathbb{R}^n$, which must be a compact set. Then the set which represents the feasible criterion points is $f(X)$, the imageof the set X under the action of f. Call this image Y.

Now construct the Pareto frontier as a subset of Y, the feasible criterion points. It can be assumed that the preferable values of each criterion parameter are the lesser ones, thus minimizing each dimension of the criterion vector. Then compare criterion vectors as follows: One criterion vector *ystrictly dominates* (or "is preferred to") a vector y* if each parameter of y is no greater than the corresponding parameter of y* and at least one parameter is strictly less: that is, $\mathbf{y}_i \leq \mathbf{y}^*_i$ for each$i$ and$\mathbf{y}_i < \mathbf{y}^*_i$for some i. This is written as $\mathbf{y} \succ \mathbf{y}^*$ to mean that y strictly dominates y*. Then the Pareto frontier is the set of points from Y that are not strictly dominated by another point in Y.

Formally, this defines a partial order on Y, namely the product order on $\mathbb{R}^m$ (more precisely, the induced order on Y as a *subset* of $\mathbb{R}^m$), and the Pareto frontier is the set of *maximal elements* with respect to this order.

Algorithms for computing the Pareto frontier of a finite set of alternatives have been studied in computer science, being sometimes referred to as the maximum vector problem or the skyline query.

Relationship to Marginal Rate of Substitution

At a Pareto efficient allocation (on the Pareto frontier), the marginal rate of substitution is the same for all consumers. A formal statement can be derived by considering a system with m consumers and $n$goods, and a utility function of each consumer as $z_i = f^i(x^i)$ where $x^i = (x_1^i, x_2^i, \ldots, x_n^i)$ is the vector of goods, both for all i. The supply constraint is written $\sum_{i=1}^{m} x_j^i = b_j^0$ for $j = 1, \ldots, n$. To optimize this problem, the Lagrangian is used:

$$L(x, \lambda, \Gamma) = f^1(x^1) + \sum_{i=2}^{m} \lambda_i (z_i^0 - f^i(x^i)) + \sum_{j=4}^{n} \Gamma_j (b_j^0 - \sum_{i=1}^{m} x_j^i)$$

where λ and Ã are multipliers.

Taking the partial derivative of the Lagrangian with respect to one good, i, and then taking the partial derivative of the Lagrangian with respect to another good, j, gives the following system of equations:

$$\frac{\partial L}{\partial x_j^i} = f_{x^1}^1 - \Gamma_j^0 = 0 \quad \text{for } j = 1, \ldots, n.$$

$$\frac{\partial L}{\partial x_j^i} = -\lambda_i f_{x^i}^i - \Gamma_j^0 = 0 \quad \text{for } i = 1, \ldots, m \text{ and } j = 1, \ldots, m,$$

where f_x is the marginal utility on f' *of* x *(the partial derivative of f with respect to* x*).*

Chapter 3

Type of Swarming

Swarm Behaviour

Swarm behaviour, or swarming, is a collective behaviour exhibited by animals of similar size which aggregate together, perhaps milling about the same spot or perhaps moving *en masse* or migrating in some direction. As a term, *swarming*is applied particularly to insects, but can also be applied to any other animal that exhibits swarm behaviour. The term *flocking* is usually used to refer specifically to swarm behaviour in birds, *herding* to refer to swarm behaviour in quadrupeds, *shoaling* or *schooling* to refer to swarm behaviour in fish.Phytoplankton also gather in huge swarms called *blooms*, although these organisms are algae and are not self propelled the way animals are. By extension, the term swarm is applied also to inanimate entities which exhibit parallel behaviours, as in a robot swarm, an earthquake swarm, or a swarm of stars.

From a more abstract point of view, swarm behaviour is the collective motion of a large number of self-propelled entities. From the perspective of the mathematical modeller, it is an emergent behaviour arising from simple rules that are followed by individuals and does not involve any central coordination.

Swarm behaviour was first simulated on a computer in 1986 with the simulation program boids. This program simulates simple agents (boids) that are allowed to move according to a set of basic rules. The model was originally designed to mimic the flocking behaviour of birds, but it can be applied also to schooling fish and other swarming entities.

Models

The simplest mathematical models of animal swarms generally represent individual animals as following three rules:

1. Move in the same direction as your neighbours
2. Remain close to your neighbours
3. Avoid collisions with your neighbours.

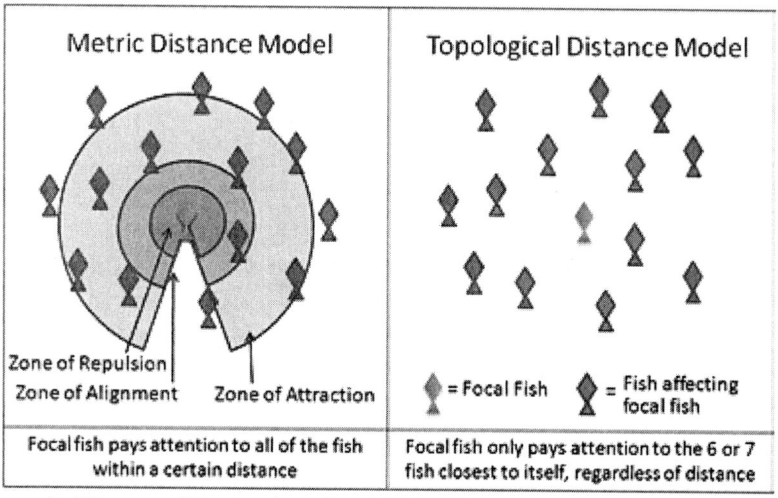

Figure: A diagram illustrating the difference between metric distance and topological distance in reference to

The boids computer program, created by Craig Reynolds in 1986, simulates swarm behaviour following the above rules. Many subsequent and current models use variations on these rules, often implementing them by means of concentric "zones" around each animal. In the *zone of repulsion*, very close to the animal, the focal animal will seek to distance itself from its neighbours to avoid collision. Slightly further away, in the *zone of alignment*, the focal animal will seek to align its direction of motion with its neighbours. In the outermost *zone of attraction*, which extends as far away from the focal animal as it is able to sense, the focal animal will seek to move towards a neighbour.

The shape of these zones will necessarily be affected by the sensory capabilities of the given animal. For example the visual field of a bird does not extend behind its body. Fish rely on both vision and onhydrodynamic perceptions relayed through their lateral line, while Antarctic krill rely both on vision and hydrodynamic signals relayed through antennae.

However recent studies of starling flocks have shown that each bird modifies its position, relative to the six or seven animals directly surrounding it, no matter how close or how far away those animals are.Interactions between flocking starlings are thus based on a topological rule rather than a metric rule. It remains to be seen whether this applies to other animals.

Swarm intelligence: Swarm intelligence is the collective behaviour ofdecentralized, self-organized systems, natural or artificial. The concept is employed in work on artificial intelligence. The expression was introduced by Gerardo Beni and Jing Wang in 1989, in the context of cellular robotic systems.

Swarm intelligence systems are typically made up of a population of simple agents or boids interacting locally with one another and with their environment. The agents follow very simple rules, and although there is no centralized control structure dictating how individual agents should behave, local, and to a certain degree random, interactions between such agents lead to theemergence of intelligent global behaviour, unknown to the individual agents.

Swarm intelligence research is multidisciplinary. It can be divided into natural swarm research studying biological systems and artificial swarm research studying human artefacts. There is also a scientific stream attempting to model the swarm systems themselves and understand their underlying mechanisms, and an engineering stream focused on applying the insights developed by the scientific stream to solve practical problems in other areas.

Emergence

The concept of emergence—that the properties and functions found at a hierarchical level are not present and are irrelevant at the lower levels–is often a basic principle behind self-organizing systems. An example of self-organization in biology leading to emergence in the natural world occurs in ant colonies.

The queen does not give direct orders and does not tell the ants what to do. Instead, each ant reacts to stimuli in the form of chemical scent from larvae, other ants, intruders, food and build up of waste, and leaves behind a chemical trail, which, in turn, provides a stimulus to other ants. Here each ant is an autonomous unit that reacts depending only on its local environment and the genetically encoded rules for its variety of ant. Despite the lack of centralized decision making, ant colonies exhibit complex behaviour and have even been able to demonstrate the ability to solve geometric problems. For example, colonies routinely find the maximum distance from all colony entrances to dispose of dead bodies.

Stigmergy

A further key concept in the field of swarm intelligence is stigmergy. Stigmergy is a mechanism of indirect coordination between agents or actions. The principle is that the trace left in the environment by an action stimulates the performance of a next action, by the same or a different agent. In that way, subsequent actions tend to reinforce and build on each other, leading to the spontaneous emergence of coherent, apparently systematic activity. Stigmergy is a form of self-organization.

It produces complex, seemingly intelligent structures, without need for any planning, control, or even direct communication between the agents. As such it supports efficient collaboration between extremely simple agents, who lack any memory, intelligence or even awareness of each other.

Algorithms

Swarm algorithms follow a Lagrangian approach or an Eulerian approach. The Eulerian approach views the swarm as a field, working with the density of the swarm and deriving mean field properties. It is a hydrodynamic approach, and can be useful for modelling the overall dynamics of large swarms. However, most models work with the Lagrangian approach, which is an agent-based model following the individual agents (points or particles) that make up the swarm. Individual particle models can follow information on heading and spacing that is lost in the Eulerian approach.

Ant Colony Optimization

Ant colony optimization is a widely used algorithm which was inspired by the behaviours of ants, and has been effective solving discrete optimization problems related to swarming. The algorithm was initially proposed by Marco Dorigo in 1992, and has since been diversified to solve a wider class of numerical problems. Species that have multiple queens may have a queen leaving the nest along with some workers to found a colony at a new site, a process akin to swarming in honeybees.

- Ants are behaviourally unsophisticated; collectively they perform complex tasks. Ants have highly developed sophisticated sign-based communication.
- Ants communicate using pheromones; trails are laid that can be followed by other ants.

- Routing problem ants drop different pheromones used to compute the "shortest" path from source to destination(s).
- Rauch EM, Millonas MM and Chialvo DR (1995) "Pattern formation and functionality in swarm models"] *Physics Letters A*, 207: 185. arXiv:adap-org/9507003.

Self-propelled Particles

Self-propelled particles (SPP) is a concept introduced in 1995 by Vicsek and Couzin et al as a special case of theboids model introduced in 1986 by Reynolds. A swarm is modelled in SPP by a collection of particles that move with a constant speed but respond to a random perturbation by adopting at each time increment the average direction of motion of the other particles in their local neighbourhood.

Simulations demonstrate that a suitable "nearest neighbour rule" eventually results in all the particles swarming together, or moving in the same direction. This emerges, even though there is no centralized coordination, and even though the neighbours for each particle constantly change over time (see the interactive simulation in the box on the right). SPP models predict that swarming animals share certain properties at the group level, regardless of the type of animals in the swarm. Swarming systems give rise to emergent behaviours which occur at many different scales, some of which are turning out to be both universal and robust. It has become a challenge in theoretical physics to find minimal statistical models that capture these behaviours.

Particle Swarm Optimization

Particle swarm optimization is another algorithm widely used to solve problems related to swarms. It was developed in 1995 by Kennedy and Eberhart and was first aimed at simulating the social behaviour and choreography of bird flocks and fish schools.

The algorithm was simplified and it was observed to be performing optimization. The system initially seeds a population with random solutions. It then searches in the problem space through successive generations using stochastic optimization to find the best solutions. The solutions it finds are called particles. Each particle stores its position as well as the best solution it has achieved so far. The particle swarm optimizer tracks thebest local value obtained so far by any particle in the local neighbourhood. The remaining particles then move through the problem space following the lead of the optimum particles. At each time iteration, the particle swarm optimiser accelerates each particle toward its optimum locations according to simple mathematical rules. Particle swarm optimization has been applied in many areas. It has few parameters to adjust, and a version that works well for a specific applications can also work well with minor modifications across a range of related applications. A book by Kennedy and Eberhart describes some philosophical aspects of particle swarm optimization applications and swarm intelligence. An extensive survey of applications is made by Poli.

Biological Swarming

Examples of biological swarming are found in bird flocks, fish schools, insect swarms,bacteria swarms, molds, molecular motors, quadruped herds and people.

Insects

The behaviour of insects that live in colonies, such as ants, bees, wasps and termites, has always been a source of fascination for children, naturalists and artists. Individual insects seems to do their own thing without any central control, yet the colony as a whole behaves in a highly coordinated manner. Researchers have found that cooperation at the colony level is largely self-organized. The group coordination that emerges is often just a consequence of the way individuals in the colony interact. These interactions can be remarkably simple, such as one ant merely following the trail left by another ant. Yet put together, the cumulative effect of such behaviours can solve highly complex problems, such as locating the shortest route in a network of possible paths to a food source. The organised behaviour that emerges in this way is sometimes called swarm intelligence.

Ants

Individual ants do not exhibit complex behaviours, yet a colony of ants collectively achieves complex tasks such as constructing nests, taking care of their young, building bridges and foraging for food. A colony of ants can locate the food source nearest their nest without any individual ant possessing any knowledge about that. If there are two paths from the ant nest to a food source, then the colony usually selects the shorter path. This is because individual ants lay down chemical trails as they move, a pheromonewhich is then followed by other ants. The ants that first return to the nest from the food source are more likely to be those that took the shorter path. More ants then retrace the shorter path, reinforcing the pheromone trail. In this way, "the ants collectively solve an optimization problem using emergent computation."

The successful techniques used by ant colonies have been studied in computer science and robotics to produce distributed and fault-tolerant systems for solving problems. This area of biomimetics has led to studies of ant locomotion, search engines that make use of "foraging trails", fault-tolerant storage and networking algorithms.

Bees

Figure: Bees swarming on a shrub

When a honey bee swarm emerges from a hive they do not fly far at first. They may gather in a tree or on a branch only a few meters from the hive. There, they cluster about the queen and send 20 -50 scout bees out to find a suitable new nest locations. The scout bees are the most experienced foragers in the cluster. An individual scout returning to the cluster promotes a location she found. She uses a dance similar to the waggle dance to indicate direction and distance to others in the cluster. The more excited she is about her findings the more excitedly she dances. If she can convince other scouts to check out the location she found, they may take off, check out the proposed site and promote the site further upon their return. Several different sites may be promoted by different scouts at first. After several hours and sometimes days, slowly a favourite location emerges from this decision making process. When all scouts agree on a final location the whole cluster takes off and flies to it. A swarm may fly for a kilometre or more to the scouted out location. This collective decision making process is remarkably successful in identifying the most suitable new nest site and keeping the swarm intact. A good nest site has to be large enough to accommodate the swarm (about 15 litres in volume), has to be well protected from the elements, receive a certain amount of warmth from the sun and be not infested with ants.

Cockroaches

Similar to ants, cockroaches leave chemical trails in their faeces as well as emitting airbornepheromones for swarming and mating. Other cockroaches will follow these trails to discover sources of food and water, and also discover where other cockroaches are hiding. Thus, cockroaches can exhibitemergent behaviour, in which group or swarm behaviour emerges from a simple set of individual interactions.

Cockroaches are mainly nocturnal and will run away when exposed to light. A study tested the hypothesis that cockroaches use just two pieces of information to decide where to go under those conditions: how dark it is and how many other cockroaches there are. The study conducted by José Halloy and colleagues at the Free University of Brussels and other European institutions created a set of tiny robots that appear to the roaches as other roaches and can thus alter the roaches' perception of critical mass. The robots were also specially scented so that they would be accepted by the real roaches.

Locusts

Figure: Swarm of desert locusts

Locusts are the swarming phase of the short-hornedgrasshoppers of the family Acrididae. Some species can breed rapidly under suitable conditions and subsequently become gregarious and migratory.

They form bands asnymphs and swarms as adults—both of which can travel great distances, rapidly stripping fields and greatlydamaging crops. The largest swarms can cover hundreds of square miles and contain billions of locusts.

A locust can eat its own weight (about 2 grams) in plants every day. That means one million locusts can eat about one ton of food each day, and the largest swarms can consume over 100,000 tonnes each day.

Swarming in locusts has been found to be associated with increased levels of serotonin which causes the locust to change colour, eat much more, become mutually attracted, and breed much more easily. Researchers propose that swarming behaviour is a response to overcrowding and studies have shown that increased tactile stimulation of the hind legs or, in some species, simply encountering other individuals causes an increase in levels of serotonin. The transformation of the locust to the swarming variety can be induced by several contacts per minute over a four-hour period.

An individual locust's response to a loss of alignment in the group appears to increase the randomness of its motion, until an aligned state is again achieved. This noise-induced alignment appears to be an intrinsic characteristic of collective coherent motion.

Insect Migration

Insect migration is the seasonal movement of insects, particularly those by species of dragonflies, beetles,butterflies and moths. The distance can vary from species to species, but in most cases these movements involve large numbers of individuals. In some cases the individuals that migrate in one direction may not return and the next generation may instead migrate in the opposite direction. This is a significant difference from bird migration.

Monarch butterflies are especially noted for their lengthy annual migration. In North America they make massive southward migrations starting in August until the first frost. A northward migration takes place in the spring. The monarch is the only butterfly that migrates both north and south as the birds do on a regular basis. But no single individual makes the entire round trip.

Figure: Cluster of monarch butterflies. Monarch butterflies migrate to Santa Cruz, Californiawhere they overwinter

Female monarchs deposit eggs for the next generation during these migrations. The length of these journeys exceeds the normal lifespan of most monarchs, which is less than two months for butterflies born in early summer. The last generation of the summer enters into a non-reproductive phase known as diapause and may live seven months or more. During diapause, butterflies fly to one of many overwintering sites. The generation that overwinters generally does not reproduce until it leaves the overwintering site sometime in February and March. It is the second, third and fourth generations that return to their northern locations in the United States and Canada in the spring. How the species manages to return to the same overwintering spots over a gap of several generations is still a subject of research; the flight patterns appear to be inherited, based on a combination of the position of the sun in the sky and a time-compensated Sun compass that depends upon acircadian clock that is based in their antennae.

Birds

Approximately 1800 of world's 10,000 bird species are long-distance migrants. The primary motivation for migration appears to be food; for example, some hummingbirds choose not to migrate if fed through the winter. Also, the longer days of the northern summer provide extended time for breeding birds to feed their young. This helps diurnalbirds to produce larger clutches than related non-migratory species that remain in the tropics. As the days shorten in autumn, the birds return to warmer regions where the available food supply varies little with the season. These advantages offset the high stress, physical exertion costs, and other risks of the migration such as predation.

Many, if not most, birds migrate in flocks. For larger birds, flying in flocks reduces energy costs. Geese in a V-formation may conserve 12–20% of the energy they would need to fly alone. Red knots and dunlins were found in radar studies to fly 5 km per hour faster in flocks than when they were flying alone. Flocks of birds save energy when flying together, much in the way that bicyclists draft one another in a peloton. Geese flying in a Vee formation save energy by flying in the updraft of the wingtip vortex generated by the previous animal in the formation. Thus, the birds flying behind do not need to work as hard to achieve lift. Studies show that birds in a V formation place themselves roughly at the optimum distance predicted by simple aerodynamic theory.

The V formation greatly boosts the efficiency and range of flying birds, particularly over long migratoryroutes. All the birds except the first fly in the upwash from the wingtip vortices of the bird ahead. The upwash assists each bird in supporting its own weight in flight, in the same way a glider can climb or maintain height indefinitely in rising air. In a V formation of 25 members, each bird can achieve a reduction of induced drag by up to 65% and as a result increase their range by 71%. The birds flying at the tips and at the front are rotated in a timely cyclical fashion to spread flight fatigue equally among the flock members. The formation also makes communication easier and allows the birds to maintain visual contact with each other.

Other animals may use similar drafting techniques when migrating. Lobsters, for example, migrate in close single-file formation "lobster trains", sometimes for hundreds of miles.

Some bar-tailed godwits have the longest known non-stop flight of any migrant, flying 11,000 km from Alaska to theirNew Zealand non-breeding areas. Prior to migration, 55 percent of their bodyweight is stored fat to fuel this uninterrupted journey.

The Mediterranean and other seas present a major obstacle to soaring birds, which must cross at the narrowest points. Massive numbers of large raptors and storks pass through areas such as Gibraltar,Falsterbo, and the Bosphorus at migration times. More common species, such as the Honey Buzzard, can be counted in hundreds of thousands in autumn. Other barriers, such as mountain ranges, can also cause funnelling, particularly of large diurnal migrants. This is a notable factor in theCentral American migratory bottleneck. This concentration of birds during migration can put species at risk. Some spectacular migrants have already gone extinct, the most notable being the Passenger Pigeon. During migration the flocks were a mile (1.6 km) wide and 300 miles (500 km) long, taking several days to pass and containing up to a billion birds.

Marine Life

Fish

The term "shoal" can be used to describe any group of fish, including mixed-species groups, while "school" is used for more closely knit groups of the same species swimming in a highly synchronised and polarised manner.

Fish derive many benefits from shoaling behaviour including defence against predators (through better predator detection and by diluting the chance of capture), enhancedforaging success, and higher success in finding a mate.It is also likely that fish benefit from shoal membership through increased hydrodynamic efficiency.

Fish use many traits to choose shoalmates. Generally they prefer larger shoals, shoalmates of their own species, shoalmates similar in size and appearance to themselves, healthy fish, and kin (when recognised). The "oddity effect" posits that any shoal member that stands out in appearance will be preferentially targeted by predators. This may explain why fish prefer to shoal with individuals that resemble them. The oddity effect would thus tend to homogenise shoals.

One puzzling aspect of shoal selection is how a fish can choose to join a shoal of animals similar to themselves, given that it cannot know its own appearance. Experiments with zebrafish have shown that shoal preference is a learned ability, not innate. A zebrafish tends to associate with shoals that resemble shoals in which it was reared, a form of imprinting.

Other open questions of shoaling behaviour include identifying which individuals are responsible for the direction of shoal movement.

In the case of migratory movement, most members of a shoal seem to know where they are going. In the case of foraging behaviour, captive shoals of golden shiner (a kind ofminnow) are led by a small number of experienced individuals who knew when and where food was available.

Radakov estimated herring schools in the North Atlantic can occupy up to 4.8 cubic kilometres with fish densities between 0.5 and 1.0 fish/cubic metre. That's several billion fish in one school.

Between May and July huge numbers of sardines spawn in the cool waters of the Agulhas Bank and then follow a current of cold water northward along the east coast of South Africa. This great migration, called the sardine run, creates spectacular feeding frenzies along the coastline as marine predators, such as dolphins, sharks and gannets attack the schools.

Krill

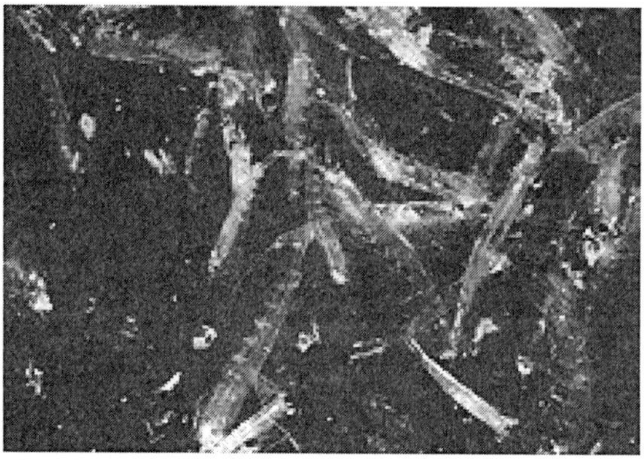

Figure: Swarming krill

Most krill, small shrimp-like crustaceans, form large swarms, sometimes reaching densities of 10,000–60,000 individual animals per cubic metre. Swarming is a defensive mechanism, confusing smaller predators that would like to pick out single individuals. Krill typically follow a diurnal vertical migration. Until recently it has been assumed that they spend the day at greater depths and rise during the night toward the surface. It has been found that the deeper they go, the more they reduce their activity, apparently to reduce encounters with predators and to conserve energy. Later work suggested that swimming activity in krill varied with stomach fullness. Satiated animals that had been feeding at the surface swim less actively and therefor sink below the mixed layer. As they sink they produce faeces which may mean that they have an important role to play in the Antarctic carbon cycle. Krill with empty stomachs were found to swim more actively and thus head towards the surface. This implies that vertical migration may be a bi- or tri-daily occurrence. Some species form surface swarms during the day for feeding and reproductive purposes even though such behaviour is dangerous because it makes them extremely vulnerable to predators. Dense swarms may elicit a feeding frenzy among fish, birds and mammal predators, especially near the surface. When disturbed, a swarm scatters, and some individuals have even been observed to moult instantaneously, leaving theexuvia behind as a decoy.

Copepods

Copepods are a group of tiny crustaceans found in the sea and lakes. Many species are planktonic (drifting in sea waters), and others are benthic (living on the ocean floor). Copepods are typically 1 to 2 millimetres (0.04 to 0.08 in) long, with a teardrop shaped body and large antennae. Although like other crustaceans they have an armouredexoskeleton, they are so small that in most species this thin armour, and the entire body, is almost totally transparent. Copepods have a compound, median single eye, usually bright red, in the centre of the transparent head.

Copepods also swarm. For example, monospecific swarms have been observed regularly around coral reefs and sea grass, and in lakes. Swarms densities were about one million copepods per cubic metre. Typical swarms were one or two metres in diameter, but some exceeded 30 cubic metres. Copepods need visual contact to keep together, and they disperse at night.

Spring produces blooms of swarming phytoplankton which provide food for copepods. Planktonic copepods are usually the dominant members of the zooplankton, and are in turn major food organisms for many other marine animals. In particular, copepods are prey to forage fish and jellyfish, both of which can assemble in vast, million-strong swarms. Some copepods have extremely fast escape responses when a predator is sensed and can jump with high speed over a few millimetres (see animated image below).

Planktonic copepods are important to the carbon cycle. Some scientists say they form the largest animal biomass on earth. They compete for this title with Antarctic krill. Because of their smaller size and relatively faster growth rates, however, and because they are more evenly distributed throughout more of the world's oceans, copepods almost certainly contribute far more to thesecondary productivity of the world's oceans, and to the global ocean carbon sink than krill, and perhaps more than all other groups of organisms together. The surface layers of the oceans are currently believed to be the world's largest carbon sink, absorbing about 2 billion tons of carbon a year, the equivalent to perhaps a third of human carbon emissions, thus reducing their impact. Many planktonic copepods feed near the surface at night, then sink into deeper water during the day to avoid visual predators. Their moulted exoskeletons, faecal pellets and respiration at depth all bring carbon to the deep sea.

Algal Blooms

Many single-celled organisms called phytoplankton live in oceans and lakes. When certain conditions are present, such as high nutrient or light levels, these organisms reproduce explosively. The resulting dense swarm of phytoplankton is called an algal bloom. Blooms can cover hundreds of square kilometres and are easily seen in satellite images. Individual phytoplankton rarely live more than a few days, but blooms can last weeks.

Other Animals

Bacteria

Swarming is also used to describe groupings of some kinds of bacteria such as myxobacteria. Myxobacteria swarm together in "wolf packs", actively moving using a process known as bacterial gliding and keeping together with the help of intercellular molecular [[signal trC. Allison and C. Hughes. Bacterial swarming: an example of procaryotic differentiation and multicellular behaviour.

Quadrupeds

Figure: Sheep dogs control the flocking behaviour of sheep

People

A collection of people can also exhibit swarm behaviour, such as pedestrians or soldiers swarming the parapets. In Cologne, Germany, two biologists from the University of Leeds demonstrated flock like behaviour in humans. The group of people exhibited similar behavioural pattern to a flock, where if five percent of the flock changed direction the others would follow. If one person was designated as a predator and everyone else was to avoid him, the flock behaved very much like a school of fish.Understanding how humans interact in crowds is important if crowd management is to effectively avoid casualties at football grounds, music concerts and subway stations.

The mathematical modelling of flocking behaviour is a common technology, and has found uses in animation. Flocking simulations have been used in many films togenerate crowds which move realistically. Tim Burton's*Batman Returns* was the first movie to make use of swarm technology for rendering, realistically depicting the movements of a group of bats using the boidssystem. The *Lord of the Rings* film trilogy made use of similar technology, known as massive, during battle scenes. Swarm technology is particularly attractive because it is cheap, robust, and simple.

An ant-based computer simulation using only six interaction rules has also been used to evaluate aircraft boarding behaviour.

Also, ants provided some guidance for Douglas A. Lawson, a systems analyst at Southwest Airlines.:"Because we know that ants have accomplished these amazing things right based on very simple rules, we know that if we want to see something complicated happen - like completely filling the interior of an aircraft with people - we know that simple ant-type behavior is adequate to represent what's occurring," Lawson said. "So Southwest Airlines said, 'Help us figure out the most efficient way to help us get our passengers on a plane,' and you said, 'I know - I'll use ants'?" asked Salie. "Yeah, right. Because they do complicated things with very simple rules," Lawson said. Lawson used mathematically-modeled ants to determine the most efficient way of boarding a plane, which turns out to be open seating. "So Southwest's way of boarding without seat numbers is actually more efficient than when I board another airline and know exactly what my seat is?" asked Salie. "Right. When we simulated what the different airlines are doing, it turns out that with assigned seats, there's a one-third chance that you're going to ask two people to get up, whereas open seating - since the middle seat is the undesirable one - generally that's the one that's last to be filled, [so] only one person is likely to get up, the person sitting near the aisle," said Lawson. "I may have to ask somebody to get up and get out of the way to let me get to a seat, and that's about it. So it's really simple. "So the ants are sort of an analogy - simple rules produce complicated transactions and events and major structures. and we can do the same thing in a little simulated world and represent very complicated processes," said Lawson.

Airlines have also used ant-based routing in assigning aircraft arrivals to airport gates. An airline system developed by Douglas A. Lawson uses swarm theory, or swarm intelligence – the idea that a colony of ants works better than one alone. Each pilot acts like an ant searching for the best airport gate. "The pilot learns from his experience what's the best for him, and it turns out that that's the best solution for the airline," Lawson explains. As a result, the "colony" of pilots always go to gates they can arrive and depart quickly. The program can even alert a pilot of plane back-ups before they happen. "We can anticipate that it's going to happen, so we'll have a gate available," says Lawson.

Swarm behaviour also occurs in traffic flow dynamics, such as the traffic wave. Bidirectional traffic can be observed in ant trails. In recent years this behaviour has been researched for insight into pedestrian and traffic models. Simulations based on pedestrian models have also been applied crowds which stampede because of panic.

Herd behaviour in marketing is used to explain the dependencies of customers' mutual behaviour. *The Economist* reported a recent conference in Rome on the subject of the simulation of adaptive human behaviour. It shared mechanisms to increase impulse buying and get people "to buy more by playing on the herd instinct." The basic idea is that people will buy more of products that are seen to be popular, and several feedback mechanisms to get product popularity information to consumers are mentioned, including smart card technology and the use of Radio Frequency Identification Tagtechnology. A "swarm-moves" model was introduced by a Florida Institute of Technology researcher, which is appealing to supermarkets because it can "increase sales without the need to give people discounts."

The application of swarm principles to robots is calledswarm robotics, while swarm intelligence refers to the more general set of algorithms.

Inspired by colonies of insects such as ants and bees, researchers are modelling the behaviour of swarms of thousands of tiny robots which together perform a useful task, such as finding something hidden, cleaning, or spying. Each robot is quite simple, but the emergent behaviour of the swarm is more complex. The whole set of robots can be considered as one single distributed system, in the same way an ant colony can be considered asuperorganism, exhibiting swarm intelligence. The largest swarms so far created include the iRobot swarm, the SRI/MobileRobots CentiBots project and the Open-source Micro-robotic Project swarm, which are being used to research collective behaviours. Swarms are also more resistant to failure. Whereas one large robot may fail and ruin a mission, a swarm can continue even if several robots fail. This could make them attractive for space exploration missions, where failure is normally extremely costly.

The I-Swarm project aims to develop a large scale swarm of up to 1,000 small microrobots, about 2 x 2 x 1 mm in size. The small size means that sensory and computational capabilities will be limited. This will be compensated for by the emergent swarm effects. The robots operate on solar power and can communicate between each other. The robots will differ in the type of sensors, manipulators and the amount of computational power. The swarm is expected to have many applications, including micro assembly, biological, medical or cleaning tasks.

Military

Military swarming is a behaviour where autonomous or partially autonomous units of action attack an enemy from several different directions and then regroup. *Pulsing*, where the units shift the point of attack, is also a part of military swarming. Military swarming involves the use of a decentralized force against an opponent, in a manner that emphasizes mobility, communication, unit autonomy and coordination or synchronization. Historically military forces used principles of swarming without really examining them explicitly, but now active research consciously examines military doctrines that draw ideas from swarming.

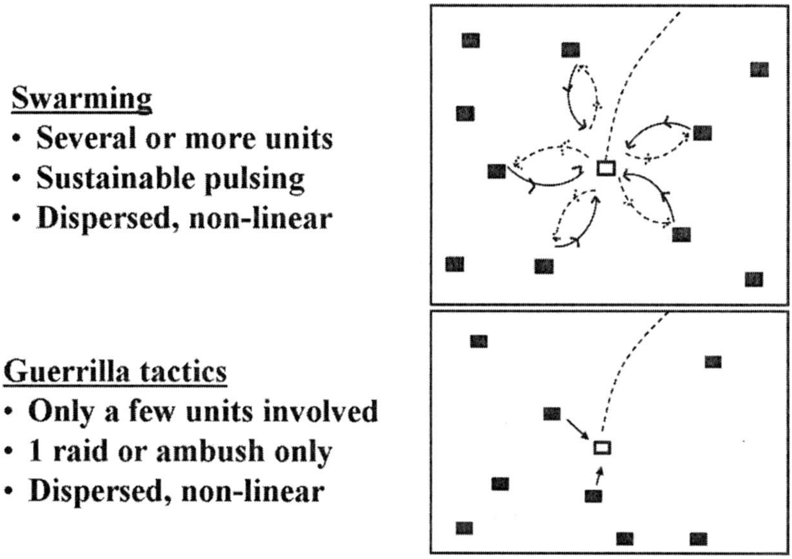

Figure: Contrast between guerilla ambush and true swarming

Merely because multiple units converge on a target, they are not necessarily swarming. Siege operations do not involve swarming, because there is no manoeuvre; there is convergence but on the besieged fortification. Nor do guerilla ambushes constitute swarms, because they are "hit-and-run". Even though the ambush may have several points of attack on the enemy, the guerillas withdraw when they either have inflicted adequate damage, or when they are endangered.

Myths

Figure: Piranha are rather fearful fish, who school for protection from predators.

- There is a popular myth that lemmings commit mass suicide by swarming off cliffs when they migrate. Driven by strong biological urges, some species of lemmings may migrate in large groups when population density becomes too great. Lemmings can swim and may choose to cross a body of water in search of a new habitat. In such cases, many may drown if the body of water is so wide as to stretch their physical capability to the limit. This fact combined with an unexplained fluctuations in the population of Norwegian lemmings gave rise to the myth.
- Piranha have a reputation as fearless fish that swarm in ferocious and predatory packs. However, recent research, which started "with the premise that they school as a means of cooperative hunting", discovered that they were in fact rather fearful fish, like other fish, who schooled for protection from their predators, such as cormorants, caimans and dolphins. Piranhas are "basically like regular fish with large teeth".

Agent-based Model

An agent-based model (ABM) (also sometimes related to the term multi-agent system or multi-agent simulation) is a class of computational models for simulating the actions and interactions of autonomous agents (both individual or collective entities such as organizations or groups) with a view to assessing their effects on the system as a whole. It combines elements of game theory, complex systems, emergence, computational sociology, multi-agent systems, and evolutionary programming.Monte Carlo Methods are used to introduce randomness. ABMs are also called individual-based models. A review of recent literature on individual-based models, agent-based models and multiagent systems is given in

The models simulate the simultaneous operations and interactions of multiple agents, in an attempt to re-create and predict the appearance of complex phenomena. The process is one of emergence from the lower (micro) level of systems to a higher (macro) level. As such, a key notion is that simple behavioral rules generate complex behavior. This principle, known as K.I.S.S. ("Keep it simple and short") is extensively adopted in the modeling community. Another central tenet is that the whole is greater than the sum of the parts. Individual agents are typically characterized as boundedly rational, presumed to be acting in what they perceive as their own interests, such as reproduction, economic benefit, or social status, using heuristics or simple decision-making rules. ABM agents may experience "learning", adaptation, and reproduction.

Most agent-based models are composed of: (1) numerous agents specified at various scales (typically referred to as agent-granularity); (2) decision-making heuristics; (3) learning rules or adaptive processes; (4) an interaction topology; and (5) a non-agent environment.

History

The idea of agent-based modeling was developed as a relatively simple concept in the late 1940s. Since it requires computation-intensive procedures, it did not become widespread until the 1990s.

The history of the agent-based model can be traced back to the Von Neumann machine, a theoretical machine capable of reproduction. The device von Neumann proposed would follow precisely detailed instructions to fashion a copy of itself. The concept was then improved by von Neumann's friendStanis³aw Ulam, also a mathematician; Ulam suggested that the machine be built on paper, as a collection of cells on a grid. The idea intrigued von Neumann, who drew it up—creating the first of the devices later termed cellular automata.

Another improvement was introduced by the mathematician John Conway. He constructed the well-known Game of Life. Unlike von Neumann's machine, Conway's Game of Life operated by tremendously simple rules in a virtual world in the form of a 2-dimensional checkerboard.

One of the earliest agent-based models in concept was Thomas Schelling's segregation model, which was discussed in his paper *Dynamic Models of Segregation* in 1971. Though Schelling originally used coins and graph paper rather than computers, his models embodied the basic concept of agent-based models as autonomous agents interacting in a shared environment with an observed aggregate, emergent outcome.

In the early 1980s, Robert Axelrod hosted a tournament of Prisoner's Dilemma strategies and had them interact in an agent-based manner to determine a winner. Axelrod would go on to develop many other agent-based models in the field of political science that examine phenomena from ethnocentrismto the dissemination of culture (Axelrod 1997). In the late 1980s, Craig Reynolds' work on flocking models contributed to the development of some of the first biological agent-based models that contained social characteristics. He tried to model the reality of lively biological agents, known as artificial life, a term coined by Christopher Langton.

The first use of the word "agent" and a definition as it is currently used today is hard to track down. One candidate appears to be John Holland and John H. Miller's 1991 paper "Artificial Adaptive Agents in Economic Theory" which is based on an earlier conference presentation of theirs.

At the same time, during the 1980s, social scientists, mathematicians, operations researchers, and a scattering of people from other disciplines developed Computational and Mathematical Organization Theory (CMOT). This field grew as a special interest group of The Institute of Management Sciences (TIMS) and its sister society, the Operations Research Society of America (ORSA). Through the mid-1990s, the field focused on such issues as designing effective teams, understanding the communication required for organizational effectiveness, and the behavior of social networks. With the appearance of StarLogo in 1990, SWARM and NetLogo in the mid-1990s and RePast and AnyLogic in 2000, as well as some custom-designed code, CMOT—later renamed Computational Analysis of Social and Organizational Systems (CASOS) — incorporated more and more agent-based modeling. Samuelson (2000) is a good brief overview of the early history, and Samuelson (2005) and Samuelson and Macal (2006) trace the more recent developments. Bonabeau (2002) is a good survey of the potential of agent-based modeling as of the time that its modelling software became widely available.

Kathleen M. Carley developed an early ABM, Construct , to explore the co-evolution of social networks and culture.

Joshua M. Epstein and Robert Axtell developed a large-scale ABM, the Sugarscape, to simulate and explore the role of social phenomenon such as seasonal migrations, pollution, sexual reproduction, combat, and transmission of disease and even culture.

Nigel Gilbert published the first textbook on Social Simulation: Simulation for the social scientist (1999) and established its most relevant journal: the Journal of Artificial Societies and Social Simulation.

In the late 1990s, the merger of TIMS and ORSA to form INFORMS, and the move by INFORMS from two meetings each year to one, helped to spur the CMOT group to form a separate society, the North American Association for Computational Social and Organizational Sciences (NAACSOS). Kathleen Carley, of Carnegie Mellon University, was a major contributor, especially to models of social networks, obtaining National Science Foundation funding for the annual conference and serving as the first President of NAACSOS. She was succeeded by David Sallach of the University of Chicago andArgonne National Laboratory, and then by Michael Prietula of Emory University. At about the same time NAACSOS began, the European Social Simulation Association (ESSA) and the Pacific Asian Association for Agent-Based Approach in Social Systems Science (PAAA), counterparts of NAACSOS, were organized. Nowadays, these three organizations collaborate internationally. The First World Congress on Social Simulation was held under their joint sponsorship in Kyoto, Japan, in August 2006. The Second World Congress was held in the northern Virginia suburbs of Washington, D.C., in July 2008, with George Mason University taking the lead role in local arrangements.

More recently, Ron Sun developed methods for basing agent-based simulation on models of human cognition, known as cognitive social simulation (see Sun 2006). Bill McKelvey, Suzanne Lohmann, Dario Nardi, Dwight Read and others at UCLA have also made significant contributions in organizational behavior and decision-making. Since 2001, UCLA has arranged a conference at Lake Arrowhead, California, that has become another major gathering point for practitioners in this field.

Theory

Most computational modeling research describes systems in equilibrium or as moving between equilibria. Agent-based modeling, however, using simple rules, can result in different sorts of complex and interesting behavior.

The three ideas central to agent-based models are agents as objects, emergence, and complexity.

Agent-based models consist of dynamically interacting rule-based agents. The systems within which they interact can create real-world-like complexity. These agents are:

- Intelligent and purposeful.
- Situated in space and time. They reside in networks and in lattice-like neighborhoods. The location of the agents and their

> responsive and purposeful behavior are encoded in algorithmic form in computer programs. The modeling process is best described as inductive. The modeler makes those assumptions thought most relevant to the situation at hand and then watches phenomena emerge from the agents' interactions. Sometimes that result is an equilibrium. Sometimes it is an emergent pattern. Sometimes, however, it is an unintelligible mangle.

In some ways, agent-based models complement traditional analytic methods. Where analytic methods enable humans to characterize the equilibria of a system, agent-based models allow the possibility of generating those equilibria. This generative contribution may be the most mainstream of the potential benefits of agent-based modeling. Agent-based models can explain the emergence of higher-order patterns—network structures of terrorist organizations and the Internet, power-law distributions in the sizes of traffic jams, wars, and stock-market crashes, and social segregation that persists despite populations of tolerant people. Agent-based models also can be used to identify lever points, defined as moments in time in which interventions have extreme consequences, and to distinguish among types of path dependency.

Rather than focusing on stable states, the models consider a system's robustness—the ways that complex systems adapt to internal and external pressures so as to maintain their functionalities. The task of harnessing that complexity requires consideration of the agents themselves—their diversity, connectedness, and level of interactions.

Framework

Recent work on the Modeling and simulation of Complex Adaptive Systems has demonstrated the need for combining agent-based and complex network based models. describes a framework consisting of four levels of developing models of complex adaptive systems described using several example multidisciplinary case studies:

1. Complex Network Modeling Level for developing models using interaction data of various system components.
2. Exploratory Agent-based Modeling Level for developing agent-based models for assessing the feasibility of further research. This can e.g. be useful for developing proof-of-concept models such as for funding applications without requiring an extensive learning curve for the researchers.

3. Descriptive Agent-based Modeling (DREAM) for developing descriptions of agent-based models by means of using templates and complex network-based models. Building DREAM models allows model comparison across scientific disciplines.
4. Validated agent-based modeling using Virtual Overlay Multiagent system (VOMAS) for the development of verified and validated models in a formal manner.

Other methods of describing agent-based models include code templates and text-based methods such as the ODD protocol.

Applications

Agent-based models have been used since the mid-1990s to solve a variety of business and technology problems. Examples of applications include supply chain optimization and logistics, modeling of consumer behavior, including word of mouth, social network effects, distributed computing,workforce management, and portfolio management. They have also been used to analyze traffic congestion. In these and other applications, the system of interest is simulated by capturing the behavior of individual agents and their interconnections. Agent-based modeling tools can be used to test how changes in individual behaviors will affect the system's emerging overall behavior.

Other models have analyzed the spread of epidemics, the threat of biowarfare, biological applicationsincluding population dynamics , the growth and decline of ancient civilizations, evolution of ethnocentric behavior, forced displacement/migration, language choice dynamics, and biomedical applications including inflammation and the human immune system. Agent-based models have also been used for developing decision support systems such as for breast cancer.

Recently, agent based modelling and simulation has been applied to various domains such as studying the impact of publication venues by researchers in the computer science domain (journals versus conferences). In addition, ABMS has been used to simulate information delivery in ambient assisted environments. In the domain of peer-to-Peer, ad-hoc and other self-organizing and complex networks, the usefulness of agent based modeling and simulation has been shown. The use of Computer Science based Formal Specification framework coupled with Wireless sensor networks and an Agent-based simulation has recently been demonstrated in. Agent based evolutionary search or algorithm is a new research topic for solving complex optimization problems. Further details on the topic can be found in R. Sarker and

T. Ray (2010) Agent based Evolutionary Approach: An Introduction, Agent Based Evolutionary Search, Springer series in Evolutionary Learning and Optimization, Springer, pp. 1–12.

Economics

Prior to and in the wake of the financial crisis interest has grown in ABMs as possible tools for economic analysis.ABMs do not assume the economy can achieve equilibrium and "representative agents" are replaced by agents with diverse, dynamic, and interdependent behavior includingherding. ABMs take a "bottom-up" approach and can generate extremely complex and volatile simulated economies. ABMs can represent unstable systems with crashes and booms that develop out of non-linear(disproportionate) responses to proportionally small changes. A July 2010 article in *The Economist* looked at ABMs as alternatives to the DGSE models. The journal *Nature* also encouraged agent-based modeling with an editorial that suggested ABMs can do a better job of representing financial markets and other economic complexities than standard models along with an essay by J. Doyne Farmer and Duncan Foley that argued ABMs could fulfill both the desires of Keynes to represent a complex economy and of Robert Lucas to construct models based on microfoundations. Farmer and Foley pointed to progress that has been made using ABMs to model parts of an economy, but argued for the creation of a very large model that incorporates low level models.

Agent-directed Simulation

The agent-directed simulation (ADS) metaphor distinguishes between two categories, namely "Systems for Agents" and "Agents for Systems." Systems for Agents (sometimes referred to as agents systems) are systems implementing agents for the use in engineering, human and social dynamics, military applications, and others. Agents for Systems are divided in two subcategories. Agent-supported systems deal with the use of agents as a support facility to enable computer assistance in problem solving or enhancing cognitive capabilities. Agent-based systems focus on the use of agents for the generation of model behavior in a system evaluation (system studies and analyses).

Hardware

The Software described above is designed for serial von-Neumann computer architectures. This limits the speed and scalability of these systems. A recent development is the use of data-parallel algorithms on Graphics Processing Units GPUs for ABM simulation , and . The extreme memory bandwidth combined with the sheer number crunching power of multi-processor GPUs has enabled simulation of millions of agents at tens of frames per second.

Verification and Validation of Agent-based Models

Verification and validation (V&V) of simulation models is extremely important. Verification involves debugging the model to ensure it works correctly; whereas Validation ensures that you have built the right model. Verification and validation in the social sciences domain can be seen in. In Computational Economics, validation can be examined in. In, the author proposes face validation, sensitivity analysis, calibration and statistical validation. Discrete-Event Simulation Framework approach for the validation of Agent-Based systems has been proposed.

A formal approach for V&V of all agent-based models is based on building a VOMAS (Virtual Overlay Multi-Agent System), a software engineering based approach, where a virtual overlay Multi-agent system is developed alongside the agent-based model. The agents in the Multi-Agent System are able to gather data by generation of logs as well as provide run-time validation and verification support by watch agents and also agents to check any violation of invariants at run-time. These are set by the Simulation Specialist with help from the SME (Subject Matter Expert). An example of using VOMAS for Verification and Validation of a Forest Fire simulation model. VOMAS provides a formal way of Validation and Verification. If you want to develop a VOMAS, you need to start by designing VOMAS agents along with the agents in the actual simulation preferably from the start. So, in essence, by the time your simulation model is complete, you essentially can consider to have one model which contains two models:

1. An Agent Based Model of the intended system
2. An Agent Based Model of the VOMAS.

Unlike all previous work on Verification and Validation, VOMAS agents ensure that the simulations are validated in-simulation i.e. even during execution. In case of any exceptional situations, which are programmed on the directive of the Simulation Specialist (SS), the VOMAS agents can report them. In addition, the VOMAS agents can be used to log key events for the sake of debugging and subsequent analysis of simulations. In other words, VOMAS allows for a flexible use of any given technique for the sake of Verification and Validation of an Agent-based Model in any domain.

Details of Validated agent-based modeling using VOMAS along with several case studies are given in. This thesis also gives details of "Exploratory Agent-based Modeling", "Descriptive Agent-based Modeling" in addition to "Validated Agent-based Modeling" using several worked case study examples.

Ant Colony Optimization Algorithms

In computer science and operations research, the ant colony optimization algorithm (ACO) is a probabilistictechnique for solving computational problems which can be reduced to finding good paths through graphs.

This algorithm is a member of ant colony algorithmsfamily, in swarm intelligence methods, and it constitutes some metaheuristic optimizations. Initially proposed byMarco Dorigo in 1992 in his PhD thesis, the first algorithm was aiming to search for an optimal path in a graph, based on the behavior of ants seeking a path between their colony and a source of food. The original idea has since diversified to solve a wider class of numerical problems, and as a result, several problems have emerged, drawing on various aspects of the behavior of ants.

In the natural world, ants (initially) wander randomly, and upon finding food return to their colony while laying down pheromone trails. If other ants find such a path, they are likely not to keep travelling at random, but to instead follow the trail, returning and reinforcing it if they eventually find food (see Ant communication).

Over time, however, the pheromone trail starts to evaporate, thus reducing its attractive strength. The more time it takes for an ant to travel down the path and back again, the more time the pheromones have to evaporate. A short path, by comparison, gets marched over more frequently, and thus the pheromone density becomes higher on shorter paths than longer ones.

Pheromone evaporation also has the advantage of avoiding the convergence to a locally optimal solution. If there were no evaporation at all, the paths chosen by the first ants would tend to be excessively attractive to the following ones. In that case, the exploration of the solution space would be constrained.

Thus, when one ant finds a good (i.e., short) path from the colony to a food source, other ants are more likely to follow that path, and positive feedback eventually leads all the ants following a single path. The idea of the ant colony algorithm is to mimic this behavior with "simulated ants" walking around the graph representing the problem to solve.

Detailed

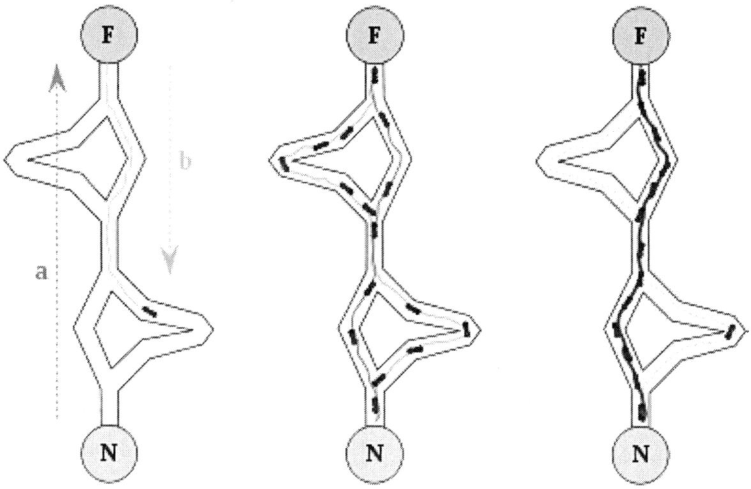

1 2 3

The original idea comes from observing the exploitation of food resources among ants, in which ants' in dividually limited cognitive abilities have collectively been able to find the shortest path between a food source and the nest.

1. The first ant finds the food source (F), via any way (a), then returns to the nest (N), leaving behind a trail pheromone (b)
2. Ants indiscriminately follow four possible ways, but the strengthening of the runway makes it more attractive as the shortest route.
3. Ants take the shortest route, long portions of other ways lose their trail pheromones.

In a series of experiments on a colony of ants with a choice between two unequal length paths leading to a source of food, biologists have observed that ants tended to use the shortest route. A model explaining this behaviour is as follows:

1. An ant (called "blitz") runs more or less at random around the colony;
2. If it d iscovers a food source, it returns more or less directly to the nest, leaving in its path a trail of pheromone;
3. These pheromones are attractive, nearby ants will be inclined to follow, more or less directly, the track;
4. Returning to the colony, these ants will strengthen the route;
5. If there are two routes to reach the same food source then, in a given amount of time, the shorter one will be traveled by more ants than the long route;
6. The short route will be increasingly enhanced, and therefore become more attractive;
7. The long route will eventually disappear because pheromones are volatile;
8. Eventually, all the ants have determined and therefore "chosen" the shortest route.

Ants use the environment as a medium of communication. They exchange information indirectly by depositing pheromones, all detailing the status of their "work". The information exchanged has a local scope, only an ant located where the pheromones were left has a notion of them. This system is called "Stigmergy" and occurs in many social animal societies (it has been studied in the case of the construction of pillars in the nests of termites). The mechanism to solve a problem too complex to be addressed by single ants is a good example of a self-organized system. This system is based on positive feedback (the deposit of pheromone attracts other ants that will strengthen it themselves) and negative (dissipation of the route by evaporation prevents the system from thrashing). Theoretically, if the quantity of pheromone remained the same over time on all edges, no route would be chosen. However, because of feedback, a slight variation on an edge will be amplified and thus allow the choice of an edge. The algorithm will move from an unstable state in which no edge is stronger than another, to a stable state where the route is composed of the strongest edges.

The basic philosophy of the algorithm involves the movement of a colony of ants through the different states of the problem influenced by two local decision policies, viz., *trails* and *attractiveness*. Thereby, each such ant incrementally constructs a solution to the problem.

When an ant completes a solution, or during the construction phase, the ant evaluates the solution and modifies the trail value on the components used in its solution. This pheromone information will direct the search of the future ants. Furthermore, the algorithm also includes two more mechanisms, viz., *trail evaporation* and*daemon actions*. *Trail evaporation* reduces all trail values over time thereby avoiding any possibilities of getting stuck in local optima. The *daemon actions* are used to bias the search process from a non-local perspective.

Common Extensions

Here are some of most popular variations of ACO Algorithms

Elitist Ant System

The global best solution deposits pheromone on every iteration along with all the other ants.

Max-Min ant System (MMAS)

Added Maximum and Minimum pheromone amounts $[\tau_{max}, \tau_{min}]$ Only global best or iteration best tour deposited pheromone. All edges are initialized to τ_{max} and reinitialized to τ_{max} when nearing stagnation.

Ant Colony System

It has been presented above.

Rank-based ant System (ASrank)

All solutions are ranked according to their length. The amount of pheromone deposited is then weighted for each solution, such that solutions with shorter paths deposit more pheromone than the solutions with longer paths.

Continuous Orthogonal ant Colony (COAC)

The pheromone deposit mechanism of COAC is to enable ants to search for solutions collaboratively and effectively. By using an orthogonal design method, ants in the feasible domain can explore their chosen regions rapidly and efficiently, with enhanced global search capability and accuracy.

The orthogonal design method and the adaptive radius adjustment method can also be extended to other optimization algorithms for delivering wider advantages in solving practical problems.

Convergence

For some versions of the algorithm, it is possible to prove that it is convergent (i.e. it is able to find the global optimum in a finite time). The first evidence of a convergence ant colony algorithm was made in 2000, the graph-based ant system algorithm, and then algorithms for ACS and MMAS.

Like mostmetaheuristics, it is very difficult to estimate the theoretical speed of convergence. In 2004, Zlochin and his colleagues showed that COA-type algorithms could be assimilated methods of stochastic gradient descent, on the cross-entropy and estimation of distribution algorithm. They proposed thesemetaheuristics as a "research-based model".

Edge Selection

An ant is a simple computational agent in the ant colony optimization algorithm. It iteratively constructs a solution for the problem at hand.

The intermediate solutions are referred to as solution states. At each iteration of the algorithm, each ant moves from a state x to state y, corresponding to a more complete intermediate solution. Thus, each ant k computes a set $A_k(x)$ of feasible expansions to its current state in each iteration, and moves to one of these in probability. For ant k, the probability p_{xy}^k of moving from state x to state y depends on the combination of two values, viz., the*attractiveness* η_{xy} of the move, as computed by some heuristic indicating the *a priori* desirability of that move and the *trail level* τ_{xy} of the move, indicating how proficient it has been in the past to make that particular move.

The *trail level* represents a posteriori indication of the desirability of that move. Trails are updated usually when all ants have completed their solution, increasing or decreasing the level of trails corresponding to moves that were part of "good" or "bad" solutions, respectively.

In general, the kth ant moves from state x to state y with probability

$$p_{xy}^k = \frac{(\tau_{xy}^{\alpha})(\eta_{xy}^{\beta})}{\sum(\tau_{xy}^{\alpha})(\eta_{xy}^{\beta})}$$

where

τ_{xy} is the amount of pheromone deposited for transition from state x to y, $0 \leq \alpha$ is a parameter to control the influence of τ_{xy}, η_{xy} is the desirability of state transition xy (*a priori* knowledge, typically1 / d_{xy}, where d is the distance) and β e" 1 is a parameter to control the influence of η_{xy}.

Pheromone Update

When all the ants have completed a solution, the trails are updated by $\tau_{xy}^{k} = (1-\rho)\tau_{xy}^{k} + \Delta\tau_{xy}^{k}$

where

τ_{xy}^{k} is the amount of pheromone deposited for a state transition xy, ρ is the *pheromone evaporation coefficient* and $\Delta\tau_{i,j}^{k}$ is the amount of pheromone deposited, typically given for a TSP problem (with moves corresponding to arcs of the graph) by

$$\Delta\tau_{xy}^{k} = \begin{cases} Q/L_k & \text{if ant } k \text{ uses curve } xy \text{ in its tour} \\ 0 & \text{otherwise} \end{cases}$$

where L_k is the cost of the kth ant's tour (typically length) and Q is a constant.

Applications

Figure: Knapsack problem. The ants prefer the smaller drop of honey over the more abundant, but less nutritious, sugar.

Ant colony optimization algorithms have been applied to many combinatorial optimization problems, ranging from quadratic assignment to protein folding or routing vehiclesand a lot of derived methods have been adapted to dynamic problems in real variables, stochastic problems, multi-targets and parallel implementations. It has also been used to produce near-optimal solutions to the travelling salesman problem. They have an advantage over simulated annealingand genetic algorithm approaches of similar problems when the graph may change dynamically; the ant colony algorithm can be run continuously and adapt to changes in real time. This is of interest in network routing and urban transportation systems. As a very good example, ant colony optimization algorithms have been used to produce near-optimal solutions to the travelling salesman problem. The first ACO algorithm was called the Ant system and it was aimed to solve the travelling salesman problem, in which the goal is to find the shortest round-trip to link a series of cities. The general algorithm is relatively simple and based on a set of ants, each making one of the possible round-trips along the cities. At each stage, the ant chooses to move from one city to another according to some rules:

1. It must visit each city exactly once;
2. A distant city has less chance of being chosen (the visibility);
3. The more intense the pheromone trail laid out on an edge between two cities, the greater the probability that that edge will be chosen;
4. Having completed its journey, the ant deposits more pheromones on all edges it traversed, if the journey is short;
5. After each iteration, trails of pheromones evaporate.

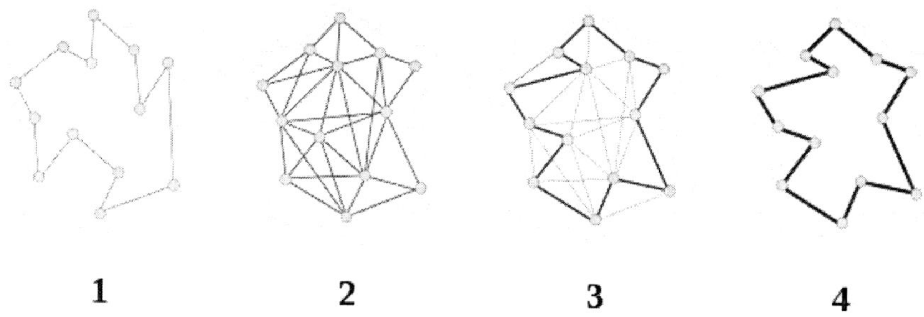

Scheduling Problem

- Job-shop scheduling problem (JSP)
- Open-shop scheduling problem (OSP)
- Permutation flow shop problem (PFSP)
- Single machine total tardiness problem (SMTTP)
- Single machine total weighted tardiness problem (SMTWTP)
- Resource-constrained project scheduling problem (RCPSP)
- Group-shop scheduling problem (GSP)
- Single-machine total tardiness problem with sequence dependent setup times (SMTTPDST)

- Multistage Flowshop Scheduling Problem (MFSP) with sequence dependent setup/changeover times.

Vehicle routing problem

- Capacitated vehicle routing problem (CVRP)
- Multi-depot vehicle routing problem (MDVRP)
- Period vehicle routing problem (PVRP)
- Split delivery vehicle routing problem (SDVRP)
- Stochastic vehicle routing problem (SVRP)
- Vehicle routing problem with pick-up and delivery (VRPPD)
- Vehicle routing problem with time windows (VRPTW)
- Time Dependent Vehicle Routing Problem with Time Windows (TDVRPTW).

Assignment Problem

- Quadratic assignment problem (QAP)
- Generalized assignment problem (GAP)
- Frequency assignment problem (FAP)
- Redundancy allocation problem (RAP).

Set Problem

- Set covering problem(SCP)
- Set partition problem (SPP)
- Weight constrained graph tree partition problem (WCGTPP)
- Arc-weighted l-cardinality tree problem (AWlCTP)
- Multiple knapsack problem (MKP)
- Maximum independent set problem (MIS).

Others

- Classification
- Connection-oriented network routing
- Connectionless network routing
- Data mining
- Discounted cash flows in project scheduling
- Distributed Information Retrieval
- Grid Workflow Scheduling Problem
- Image processing

- Intelligent testing system
- System identification
- Protein Folding
- Power Electronic Circuit Design.

Definition Difficulty

With an ACO algorithm, the shortest path in a graph, between two points A and B, is built from a combination of several paths. It is not easy to give a precise definition of what algorithm is or is not an ant colony, because the definition may vary according to the authors and uses. Broadly speaking, ant colony algorithms are regarded aspopulated metaheuristics with each solution represented by an ant moving in the search space. Ants mark the best solutions and take account of previous markings to optimize their search. They can be seen as probabilisticmulti-agent algorithms using a probability distribution to make the transition between each iteration. In their versions for combinatorial problems, they use an iterative construction of solutions. According to some authors, the thing which distinguishes ACO algorithms from other relatives (such as algorithms to estimate the distribution or particle swarm optimization) is precisely their constructive aspect. In combinatorial problems, it is possible that the best solution eventually be found, even though no ant would prove effective. Thus, in the example of the Travelling salesman problem, it is not necessary that an ant actually travels the shortest route: the shortest route can be built from the strongest segments of the best solutions. However, this definition can be problematic in the case of problems in real variables, where no structure of 'neighbours' exists. The collective behaviour of social insects remains a source of inspiration for researchers. The wide variety of algorithms (for optimization or not) seeking self-organization in biological systems has led to the concept of "swarm intelligence", which is a very general framework in which ant colony algorithms fit.

Stigmergy Algorithms

There is in practice a large number of algorithms claiming to be "ant colonies", without always sharing the general framework of optimization by canonical ant colonies (COA). In practice, the use of an exchange of information between ants via the environment (a principle called "Stigmergy") is deemed enough for an algorithm to belong to the class of ant colony algorithms. This principle has led some authors to create the term "value" to organize methods and behavior based on search of food, sorting larvae, division of labour and cooperative transportation.

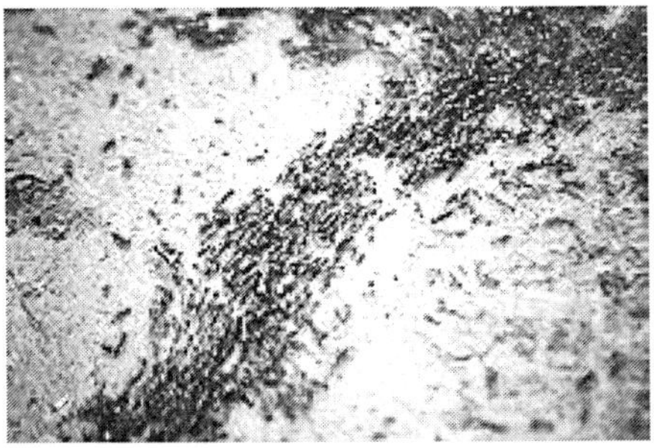

Related Methods

- Genetic algorithms (GA) maintain a pool of solutions rather than just one. The process of finding superior solutions mimics that of evolution, with solutions being combined or mutated to alter the pool of solutions, with solutions of inferior quality being discarded.
- Simulated annealing (SA) is a related global optimization technique which traverses the search space by generating neighboring solutions of the current solution. A superior neighbor is always accepted. An inferior neighbor is accepted probabilistically based on the difference in quality and a temperature parameter. The temperature parameter is modified as the algorithm progresses to alter the nature of the search.
- Reactive search optimization focuses on combining machine learning with optimization, by adding an internal feedback loop to self-tune the free parameters of an algorithm to the characteristics of the problem, of the instance, and of the local situation around the current solution.
- Tabu search (TS) is similar to simulated annealing in that both traverse the solution space by testing mutations of an individual solution. While simulated annealing generates only one mutated solution, tabu search generates many mutated solutions and moves to the solution with the lowest fitness of those generated. To prevent cycling and encourage greater movement through the solution space, a tabu list is maintained of partial or complete solutions. It is forbidden to move to a solution that contains elements of the tabu list, which is updated as the solution traverses the solution space.

- Artificial immune system (AIS) algorithms are modeled on vertebrate immune systems.
- Particle swarm optimization (PSO), a Swarm intelligence method
- Intelligent Water Drops (IWD), a swarm-based optimization algorithm based on natural water drops flowing in rivers
- Gravitational Search Algorithm (GSA), a Swarm intelligence method
- Ant colony clustering method (ACCM), a method that make use of clustering approach,extending the ACO.
- Stochastic diffusion search (SDS), an agent-based probabilistic global search and optimization technique best suited to problems where the objective function can be decomposed into multiple independent partial-functions

History

Chronology of Ant colony optimization algorithms:

- 1959, Pierre-Paul Grassé invented the theory ofStigmergy to explain the behavior of nest building intermites;
- 1983, Deneubourg and his colleagues studied thecollective behavior of ants;
- 1988, and Moyson Manderick have an article on self-organization among ants;
- 1989, the work of Goss, Aron, Deneubourg and Pasteels on the collective behavior of Argentine ants, which will give the idea of Ant colony optimization algorithms;
- 1989, implementation of a model of behavior for food by Ebling and his colleagues;
- 1991, M. Dorigo proposed the Ant System in his doctoral thesis (which was published in 1992). A technical report extracted from the thesis and co-authored by V. Maniezzo and A. Colorni was published five years later;
- 1996, publication of the article on Ant System;
- 1996, Hoos and Stützle invent the MAX-MIN Ant System;
- 1997, Dorigo and Gambardella publish the Ant Colony System;
- 1997, Schoonderwoerd and his colleagues developed the first application to telecommunicationnetworks;

- 1998, Dorigo launches first conference dedicated to the ACO algorithms;
- 1998, Stützle proposes initial parallel implementations;
- 1999, Bonabeau, Dorigo and Theraulaz publish a book dealing mainly with artificial ants
- 2000, special issue of the Future Generation Computer Systems journal on ant algorithms
- 2000, first applications to the scheduling, scheduling sequence and the satisfaction of constraints;
- 2000, Gutjahr provides the first evidence of convergence for an algorithm of ant colonies
- 2001, the first use of COA Algorithms by companies (Eurobios and AntOptima);
- 2001, IREDA and his colleagues published the first multi-objective algorithm
- 2002, first applications in the design of schedule, Bayesian networks;
- 2002, Bianchi and her colleagues suggested the first algorithm for stochastic problem;
- 2004, Zlochin and Dorigo show that some algorithms are equivalent to the stochastic gradient descent, the cross-entropy and algorithms to estimate distribution
- 2005, first applications to protein folding problems.

Ant Robotics

Ant robotics is a special case of swarm robotics. Swarm robots are simple and cheap robots with limited sensing and computational capabilities. This makes it feasible to deploy teams of swarm robots and take advantage of the resulting fault tolerance and parallelism. Swarm robots cannot use conventional planning methods due to their limited sensing and computational capabilities. Thus, their behavior is often driven by local interactions.

Ant robots are swarm robots that can communicate via markings, similar to ants that lay and follow pheromone trails. Some ant robots use long-lasting trails (either regular trails of a chemical substance or smart trails of transceivers), others use short-lasting trails (heat, odor or alcohol), and others even use virtual trails.

Invention

In 1991, American electrical engineer James McLurkin was the first to conceptualize the idea of "robot ants" while working at the MIT Computer Science and Artificial Intelligence Laboratory at the Massachusetts Institute of Technology. The robots consisted of sensors, infrared emitters, and communication systems capable of detecting objects in their path. McLurkin's invention was through studying the behavior of real ants in ant colonies and keeping ant farms as a basis for his programming. Through this examination, he could better understand how insects structured their workloads in order to produce a viable and working prototype of robotic ants.

Background

Researchers have developed ant robot hardware and software and demonstrated, both in simulation and on physical robots, that single ant robots or teams of ant robots solve robot-navigation tasks (such as path following and terrain coverage) robustly and efficiently. For example, trails coordinate the ant robots via implicit communication and provide an alternative toprobabilistic reasoning for solving the simultaneous localization and mapping problem.

Researchers have also developed a theoretical foundation for ant robotics, based on ideas from real-time heuristic search,stochastic analysis and graph theory. Recently, it was shown that a single ant robot (modeled as finite state machine) can simulate the execution of any arbitrary Turing machine. This proved that a single ant robot, using pheromones, can execute arbitrarily complex single-robot algorithms. However, the result unfortunately does not hold for N robots.

Artificial Ants

In computer science,Artificial Ants stand formulti-agent methods inspired by the behavior of real ants. The pheromone-based communication of biological ants is often the predominant paradigm used. Combinations of Artificial Ants and local search algorithms have become a method of choice for numerous optimization tasks involving some sort ofgraph, e. g., vehicle routingand internet routing. The burgeoning activity in this field has led to conferences dedicated solely to Artificial Ants, and to numerous commercial applications by specialized companies such as AntOptima. As an example, Ant colony optimization is a class ofoptimization algorithms modeled on the actions of an ant colony. Artificial 'ants' (e.g. simulation agents) locate optimal solutions by moving through a parameter space representing all possible solutions. Real ants lay down pheromones directing each other to resources while exploring their environment. The simulated 'ants' similarly record their positions and the quality of their solutions, so that in later simulation iterations more ants locate better solutions. One variation on this approach isthe bees algorithm, which is more analogous to the foraging patterns of the honey bee, another social insect.

Ambient Networks of Intelligent Objects

New concepts are required since "intelligence" is no longer centralized but can be found throughout all minuscule objects. Anthropocentric concepts have always led us to the production of IT systems in which data processing, control units and calculating forces are centralized. These centralized units have continually increased their performance and can be compared to the human brain. The model of the brain has become the ultimate vision of computers. Ambient networks of intelligent objects and, sooner or later, a new generation of information systems which are even more diffused and based on nanotechnology, will profoundly change this concept. Small devices that can be compared to insects do not dispose of a high intelligence on their own. Indeed, their intelligence can be classed as fairly limited. It is, for example, impossible to integrate a high performance calculator with the power to solve any kind of mathematical problem into a biochip that is implanted into the human body or integrated in an intelligent tag which is designed to trace commercial articles. However, once those objects are interconnected they dispose of a form of intelligence that can be compared to a colony of ants or bees. In the case of certain problems, this type of intelligence can be superior to the reasoning of a centralized system similar to the brain.

Nature has given us several examples of how minuscule organisms, if they all follow the same basic rule, can create a form of collective intelligence on the macroscopic level. Colonies of social insects perfectly illustrate this model which greatly differs from human societies. This model is based on the co-operation of independent units with simple and unpredictable behavior. They move through their surrounding area to carry out certain tasks and only possess a very limited amount of information to do so. A colony of ants, for example, represents numerous qualities that can also be applied to a network of ambient objects. Colonies of ants have a very high capacity to adapt themselves to changes in the environment as well as an enormous strength in dealing with situations where one individual fails to carry out a given task. This kind of flexibility would also be very useful for mobile networks of objects which are perpetually developing. Parcels of information that move from a computer to a digital object behave in the same way as ants would do. They move through the network and pass from one knot to the next with the objective of arriving at their final destination as quickly as possible.

Bees Algorithm

In computer science and operations research, the bees algorithm is a population-based search algorithm first developed in 2005. It mimics the food foraging behaviour of swarms of honey bees. In its basic version, the algorithm performs a kind of neighbourhood search combined with random search and can be used for both combinatorial optimization and functional optimisation.

The Foraging Process in Nature

A colony of honey bees can extend itself over long distances (up to 14 km) and in multiple directions simultaneously to exploit a large number of food sources. A colony prospers by deploying its foragers to good fields. In principle, flower patches with plentiful amounts of nectar or pollen that can be collected with less effort should be visited by more bees, whereas patches with less nectar or pollen should receive fewer bees.

The foraging process begins in a colony by scout bees being sent to search for promising flower patches. Scout bees move randomly from one patch to another. During the harvesting season, a colony continues its exploration, keeping a percentage of the population as scout bees.

When they return to the hive, those scout bees that found a patch which is rated above a certain quality threshold (measured as a combination of some constituents, such as sugar content) deposit their nectar or pollen and go to the "dance floor" to perform a dance known as the waggle dance.

This dance is essential for colony communication, and contains three pieces of information regarding a flower patch: the direction in which it will be found, its distance from the hive and its quality rating (or fitness). This information helps the colony to send its bees to flower patches precisely, without using guides or maps. Each individual's knowledge of the outside environment is gleaned solely from the waggle dance. This dance enables the colony to evaluate the relative merit of different patches according to both the quality of the food they provide and the amount of energy needed to harvest it. After waggle dancing inside the hive, the dancer (i.e. the scout bee) goes back to the flower patch with follower bees that were waiting inside the hive. More follower bees are sent to more promising patches. This allows the colony to gather food quickly and efficiently.

While harvesting from a patch, the bees monitor its food level. This is necessary to decide upon the next waggle dance when they return to the hive. If the patch is still good enough as a food source, then it will be advertised in the waggle dance and more bees will be recruited to that source.

The Bees Algorithm

The Bees Algorithm is an optimisation algorithm inspired by the natural foraging behaviour of honey bees to find the optimal solution. The algorithm requires a number of parameters to be set, namely: number of scout bees (n), number of sites selected out of n visited sites (m), number of best sites out of m selected sites (e), number of bees recruited for best e sites (nep), number of bees recruited for the other (m-e) selected sites (nsp), initial size of patches (ngh) which includes site and its neighbourhood and stopping criterion.

The pseudo code for the bees algorithm in its simplest form:

1. Initialise population with random solutions.
2. Evaluate fitness of the population.
3. While (stopping criterion not met) //Forming new population.
4. Select sites for neighbourhood search.
5. Recruit bees for selected sites (more bees for best e sites) and evaluate fitnesses.
6. Select the fittest bee from each patch.
7. Assign remaining bees to search randomly and evaluate their fitnesses.
8. End While.

In first step, the bees algorithm starts with the scout bees (n) being placed randomly in the search space. In step 2, the fitnesses of the sites visited by the scout bees are evaluated. In step 4, bees that have the highest fitnesses are chosen as "selected bees" and sites visited by them are chosen for neighbourhood search. Then, in steps 5 and 6, the algorithm conducts searches in the neighbourhood of the selected sites, assigning more bees to search near to the best e sites.

The bees can be chosen directly according to the fitnesses associated with the sites they are visiting. Alternatively, the fitness values are used to determine the probability of the bees being selected. Searches in the neighbourhood of the best e sites which represent more promising solutions are made more detailed by recruiting more bees to follow them than the other selected bees.

Together with scouting, this differential recruitment is a key operation of the Bees Algorithm. However, in step 6, for each patch only the bee with the highest fitness will be selected to form the next bee population.

In nature, there is no such a restriction. This restriction is introduced here to reduce the number of points to be explored. In step 7, the remaining bees in the population are assigned randomly around the search space scouting for new potential solutions. These steps are repeated until a stopping criterion is met. At the end of each iteration, the colony will have two parts to its new population – those that were the fittest representatives from a patch and those that have been sent out randomly.

Applications

The Bees Algorithm has found many applications in engineering, such as:

- Training neural networks for pattern recognition.
- Forming manufacturing cells.
- Scheduling jobs for a production machine.
- Solving continuous problems and engineering optimization.
- Finding multiple feasible solutions to a preliminary design problems.
- Data clustering
- Optimising the design of mechanical components.
- Multi-Objective Optimisation.
- Tuning a fuzzy logic controller for a robot gymnast.
- Computer Vision and Image Analysis.

In Job Shop Scheduling

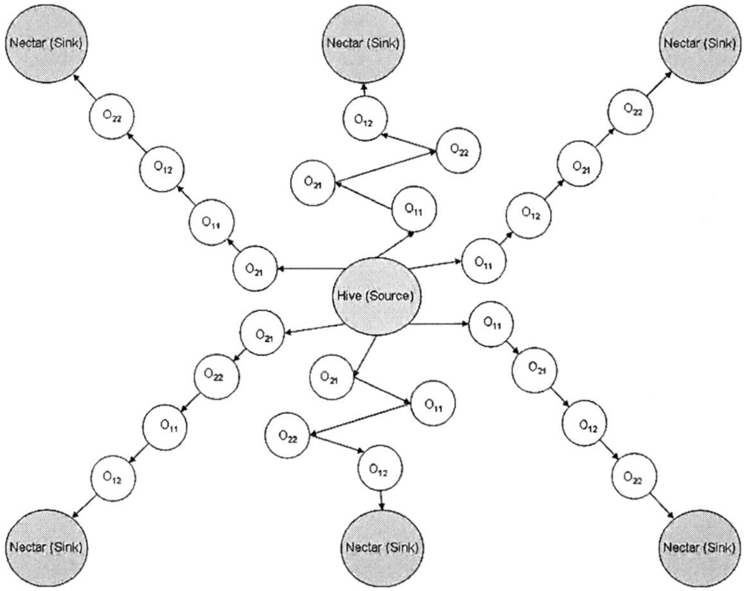

Figure: Bee colony model for job shop scheduling problems

The honey bees' effective foraging strategy can be applied to job shop scheduling problems.

A feasible solution in a job shop scheduling problem is a complete schedule of operations specified in the problem. Each solution can be thought of as a path from the hive to the food source. The figure on the right illustrates such an analogy

The makespan of the solution is analogous to the profitability of the food source in terms of distance and sweetness of the nectar. Hence, the shorter the makespan, the higher the profitability of the solution path.

We can thus maintain a colony of bees, where each bee will traverse a potential solution path. Once a feasible solution is found, each bee will return to the hive to perform a waggle dance.

The waggle dance will be represented by a list of "elite solutions", from which other bees can choose to follow another bee's path. Bees with a better makespan will have a higher probability of adding its path to the list of "elite solutions", promoting a convergence to an optimal solution.

Using the above scheme, the natural honey bee's self organizing foraging strategy can be applied to the job shop scheduling problem.

Crowd Simulation

Crowd simulation is the process of simulating the movement of a large number of objects or characters, now often appearing in 3D computer graphics for film. While simulating these crowds, observed human behavior interaction is taken into account, to replicate the collective behavior.

The need for crowd simulation arises when a scene calls for more characters than can be practically animated using conventional systems, such as skeletons/bones. Simulating crowds offer the advantages of being cost effective as well as allow for total control of each simulated character or agent.

Animators typically create a library of motions, either for the entire character or for individual body parts. To simplify processing, these animations are sometimes *baked* as morphs. Alternatively, the motions can be generated *procedurally* - i.e. choreographed automatically by software.

The actual movement and interactions of the crowd is typically done in one of two ways:

Particle Motion

The characters are attached to point particles, which are then animated by simulating wind, gravity, attractions, and collisions. The particle method is usually inexpensive to implement, and can be done in most 3D software packages. However, the method is not very realistic because it is difficult to direct individual entities when necessary, and because motion is generally limited to a flat surface.

Crowd AI

The entities - also called agents - are given artificial intelligence, which guides the entities based on one or more functions, such as sight, hearing, basic emotion, energy level, aggressiveness level, etc. The entities are given goals and then interact with each other as members of a real crowd would. They are often programmed to respond to changes in environment, enabling them to climb hills, jump over holes, scale ladders, etc. This system is much more realistic than particle motion, but is very expensive to program and implement.

The most notable examples of AI simulation can be seen in New Line Cinema's *The Lord of the Rings*films, where AI armies of many thousands battle each other. The crowd simulation was done usingWeta Digital's Massive software.

Sociology

Crowd simulation can also refer to simulations based on group dynamics and crowd psychology, often in public safety planning. In this case, the focus is just the behavior of the crowd, and not the visual realism of the simulation. Crowds have been studied as a scientific interest since the end of the 19th Century. A lot of research has focused on the collective social behavior of people at social gatherings, assemblies, protests, rebellions, concerts, sporting events and religious ceremonies. Gaining insight into natural human behavior under varying types of stressful situations will allow better models to be created which can be used to develop crowd controlling strategies.

Emergency response teams such as policemen, the National Guard, military and even volunteers must undergo some type of crowd control training. Using researched principles of human behavior in crowds can give disaster training designers more elements to incorporate to create realistic simulated disasters. Crowd behavior can be observed during panic and non-panic conditions. When natural and unnatural events toss social ideals into a twisting chaotic bind, such as the events of 9/11 and hurricane Katrina, humanity's social capabilities are truly put to the test. Military programs are looking more towards simulated training, involving emergency responses, due to their cost effective technology as well as how effective the learning can be transferred to the real world. Many events that may start out controlled can have a twisting event that turns them into catastrophic situations, where decisions need to be made on the spot. It is these situations in which crowd dynamical understanding would play a vital role in reducing the potential for anarchy.

Modeling techniques of crowds vary from holistic or network approaches to understanding individualistic or behavioral aspects of each agent. For example the Social Force Model describes a need for individuals to find a balance between social interaction and physical interaction. An approach that incorporates both aspects, and is able to adapt depending on the situation, would better describe natural human behavior, always incorporating some measure of unpredictability. With the use of multi-agent models understanding these complex behaviors has become a much more comprehensible task. With the use of this type of software, systems can now be tested under extreme conditions, and simulate conditions over long periods of time in the matter of seconds.

Firefly Algorithm

The firefly algorithm (FA) is a metaheuristic algorithm, inspired by the flashing behaviour of fireflies. The primary purpose for a firefly's flash is to act as a signal system to attract other fireflies. Xin-She Yang formulated this firefly algorithm by assuming:

1. All fireflies are unisexual, so that one firefly will be attracted to all other fireflies;
2. Attractiveness is proportional to their brightness, and for any two fireflies, the less brighter one will be attracted by (and thus move to) the brighter one; however, the brightness can decrease as their distance increases;
3. If there are no fireflies brighter than a given firefly, it will move randomly.

The brightness should be associated with the objective function. Firefly algorithm is a nature nspired metaheuristic optimization algorithm.

Algorithm Description

The pseudo code can be summarized as:

Begin

1) Objective function: $f(\mathbf{x}),\ \mathbf{x} = (x_1, x_2, \ldots, x_d)$;

2) Generate an initial population of fireflies $\mathbf{x}_i \ (i = 1, 2, \ldots, n)$

3) Formulate light intensity I so that it is associated with $f(\mathbf{x})$

(for example, for maximization problems, $I \propto f(\mathbf{x})$ or simply $I = f(\mathbf{x})$;

4) Define absorption coefficient γ

While (t<MaxGeneration)

    for i=1:n (all n fireflies)

        for j=1:i (n fireflies)

            if (I_j > I_i),

              move firefly i towards j;

            end if

        Vary attractiveness with distance r via $\exp(-\gamma\, r^2)$

```
            Evaluate new solutions and update light
intensity;
          end for j
      end for i
      Rank fireflies and find the current best;
   end while
   Post-processing the results and visualization;
```

end

The main update formula for any pair of two fireflies $\mathbf{x}_i$ and $\mathbf{x}_j$ is

$$\mathbf{x}_i^{t+1} = \mathbf{x}_i^t + \beta \exp[-\gamma r_{ij}^2] + \alpha_t \epsilon_t$$

where α_t is a parameter controlling the step size, while ϵ_t is a vector drawn from a Gaussian or other distribution.

It can be shown that the limiting case $\gamma \to 0$ corresponds to the standard Particle Swarm Optimization (PSO). In fact, if the inner loop (for j) is removed and the brightness I_j is replaced by the current global best g^*, then FA essentially becomes the standard PSO.

Implementation Guides

The γ should be related to the scales of design variables. Ideally, the β term should be order one, which requires that γ should be linked with scales. For example, one possible choice is to use $\gamma = 1/\sqrt{L}$ where L is the average scale of the problem. In case of scales vary significantly, γ can be considered as a vector to suit different scales in different dimensions. Similarly, α_t should also be linked with scales. For example, $\alpha_t \leftarrow 0.01 L \alpha_t$. It is worth pointing out the above description does not include the randomness reduction. In fact, in actual implementation by most researchers, the motion of the fireflies is gradually reduced by an annealing-like randomness reduction via $\alpha = \alpha_0 \delta^t$where $0 < \delta < 1$(*e.g.*$\delta = 0.97$). In some difficult problem, it may be helpful if you increase α_t at some stages, then reduce it when necessary. This non-monotonic variation of α_t will enable the algorithm to escape any local optima when in the unlikely case it might get stuck if randomness is reduced too quickly.

Parametric studies show that n (number of fireflies) should be about 15 to 40 for most problems.A python implementation is also available, though with limited functionalities.

Recent studies shows that the firefly algorithm is very efficient, and could outperform other metaheuristic algorithms including particle swarm optimization. Most metaheuristic algorithms may have difficulty in dealing with stochastic test functions, and it seems that firefly algorithm can deal with stochastic test functions very efficiently. In addition, FA is also better for dealing with noisy optimization problems with ease of implementation.

Variants of Firefly Algorithm

A discrete version of Firefly Algorithm, namely, Discrete Firefly Algorithm (DFA) proposed recently by M. K. Sayadi, R. Ramezanian and N. Ghaffari-Nasab can efficiently solve NP-hard scheduling problems. DFA outperforms existing algorithms such as the ant colony algorithm.

For image segmentation, the FA-based method is far more efficient to Otsu's method and recursive Otsu. Meanwhile, a good implementation of a discrete firefly algorithm for QAP problems has been carried out by Durkota.

Multiobjective FA

An important study of FA was carried out by Apostolopoulos and Vlachos, which provides a detailed background and analysis over a wide range of test problems including multobjective load dispatch problem.

Lagrangian FA

An interesting, Lagrangian firefly algorithm is proposed to solve power system optimization unit commitment problems.

Chaotic FA

A chaotic firefly algorithm (CFA) was developed and found to outperform the previously best-known solutions available.

Hybrid Algorithms

A hybrid intelligent scheme has been developed by combining the firefly algorithm with the ant colony optimization.

Applications

Digital Image Compression

Very recently, an FF-LBG algorithm for vector quantization of digital image compression was based on the firefly algorithm, which proves to be faster than other algorithms such as PSO-LBG and HBMO-LBG (particle swarm optimization and honey-bee mating optimization; variations on the Linde–Buzo–Gray algorithm). For minimum cross entropy thresholding, firefly-based algorithm uses the least computation time

Eigenvalue optimization

Eigenvalue optimization of isospectral systems has solved by FA and multiple optimum points have been found efficiently.

Feature selection and fault detection

Feature selection can be also carried out successfully using firefly algorithm.

Real-time fault identification in large systems becomes viable, based on the recent work on fault identification with binary adaptive firefly optimization.

Antenna Design

Firefly algorithms outperforms ABC for optimal design of linear array of isotropic sources.

Structural Design

For mixed-variable problems, many optimization algorithms may struggle. However, firefly algorithm can efficiently sove optimization problems with mixed variables.

Scheduling and TSP

Firefly-based algorithms for scheduling task graphs and job shop scheduling requires less computing than all other metaheuristics. A binary firefly algorithm has been developed to tackle the knapsack cryptosystem efficiently Recently, an evolutionary discrete FA has been developed for solving travelling salesman problems Further improvement in performance can be obtained by using preferential directions in firefly movements.

Semantic Web Composition

A hybrid FA has been developed by Pop et al. for selecting optimal solution in semantic web service composition.

Clustering

Performance study for clustering also suggested that firefly algorithm is very efficient.

Dynamic Problems

Firefly algorithm can solve optimization problems in dynamic environments very efficiently.

Glowworm Swarm Optimization

The glowworm swarm optimization (GSO) is a swarm intelligence optimization algorithm developed based on the behaviour of glowworms (also known as fireflies or lightning bugs). The behaviour pattern of glowworms which is used for this algorithm is the apparent capability of the glowworms to change the intensity of the luciferin emission and thus appear to glow at different intensities.

1. The GSO algorithm makes the agents glow at intensities approximately proportional to the function value being optimized. It is assumed that glowworms of brighter intensities attract glowworms that have lower intensity.
2. The second significant part of the algorithm incorporates a dynamic decision range by which the effect of distant glowworms are discounted when a glowworm has sufficient number of neighbours or the range goes beyond the range of perception of the glowworms.

The part 2 of the algorithm makes it different from Firefly algorithm where there is no "sufficient number o neighbours" limit and there is no perception limit based on distance. This two "cognitive limits" allows swarms of glowworms to split into sub-groups and converge to high function value points. This property of the algorithm allows it to be used to identify multiple peaks of a multi-modal function and makes it part of Evolutionary multi-modal optimization algorithms family.

The GSO algorithm was developed and introduced by K.N. Krishnanand and D. Ghose in 2005 at the Guidance, Control, and Decision Systems Laboratory in the Department of Aerospace Engineering at the Indian Institute of Science, Bangalore, India. Subsequently, it has been used in various applications and several papers have appeared in the literature using the GSO algorithm.

Self-propelled Particles

Self-propelled particles (SPP), also referred to as self-driven particles or as the Couzin–Vicsek algorithm,is a concept used to model swarm behaviour. The concept was introduced in 1995 by Vicsek and

Couzin *et al.*

as a special case of the Boids model introduced in 1986 byReynolds. A swarm is modelled in SPP by a collection of particles that move with a constant speed but respond to a random perturbation by adopting at each time increment the average direction of motion of the other particles in their local neighbourhood.

SPP models predict that swarming animals share certain properties at the group level, regardless of the type of animals in the swarm. Swarming systems give rise toemergent behaviours which occur at many different scales, some of which are turning out to be both universal and robust. It has become a challenge in theoretical physics to find minimal statistical models that capture these behaviours.

Overview

The SPP model is based on a collection of points or particles, each functioning individually as anautonomous agent, and each following the same simple rules which govern their behaviour. The particles move in a plane with constant speed but in different directions. The direction of each particle is updated using a "nearest neighbor rule", a local rule which replaces the direction of each particle with the average of the particle's own direction plus the directions of its immediate neighbours.

Simulations demonstrate that a suitable "nearest neighbour rule" eventually results in all the particles swarming together, or moving in the same direction. This emerges, even though there is no centralised coordination, and even though the neighbours for each particle constantly change over time (see the interactive simulation in the box on the right).

Although more realistic swarming models have been explored, the SPP model remains important because of its simplicity and the strength and the variety of its emergent phenomena. The SPP model is an agent-based model based on a Lagrangian viewpoint, which follows individual particles rather than working with the density of the swarm. It is a discrete-time switched linear system which is stable, even though no common quadratic Lyapunov function exists. It is an analogue of the Ising model in ferromagnetism, where temperature corresponds to particle randomness and spin clusters correspond to particle clusters.

Examples

Figure: Locust nymph

Marching Locusts

Young desert locusts are solitary and wingless nymphs. If food is short they can gather together and start occupying neighbouring areas, recruiting more locusts. Eventually they can become a marching army extending over many kilometres. This can be the prelude to the development of the vast flying adult locust swarms which devastate vegetation on a continental scale.

One of the key predictions of the SPP model is that as thepopulation density of a group increases, an abrupt transition occurs from individuals moving in relatively disordered and independent ways within the group to the group moving as a highly aligned whole. Thus, in the case of young desert locusts, a trigger point should occur which turns disorganised and dispersed locusts into a coordinated marching army. When the critical population density is reached, the insects should start marching together in a stable way and in the same direction.

In 2006, a group of researchers examined how this model held up in the laboratory. Locusts were placed in a circular arena, and their movements were tracked with computer software. At low densities, below 18 locusts per square metre, the locusts mill about in a disordered way. At intermediate densities, they start falling into line and marching together, punctuated by abrupt but coordinated changes in direction. However, when densities reached a critical value at about 74 locusts/ m^2, the locusts ceased making rapid and spontaneous changes in direction, and instead marched steadily in the same direction for the full eight hours of the experiment (see video on the left). This confirmed the behaviour predicted by the SPP models.

In the field, according to the Food and Agriculture Organization of the United Nations, the average density of marching bands is 50 locusts/m^2 (50 million locusts/km^2), with a typical range from 20 to

120 locusts/m^2.:[29] The research findings discussed above demonstrate the dynamic instability that is present at the lower locust densities typical in the field, where marching groups randomly switch direction without any external perturbation. Understanding this phenomenon, together with the switch to fully coordinated marching at higher densities, is essential if the swarming of desert locusts is to be controlled.

Bird Landings

Figure: Flocks of birds can abruptly change their direction in unison, and then, just as suddenly, make a unanimous group decision to land

Swarming animals, such as ants, bees, fish and birds, are often observed suddenly switching from one state to another. For example, birds abruptly switch from a flying state to a landing state. Or fish switch from schooling in one direction to schooling in another direction. Such state switches can occur with astonishing speed and synchronicity, as though all the members in the group made a unanimous decision at the same moment. Phenomena like these have long puzzled researchers.

In 2010, Bhattacharya and Vicsek used an SPP model to analyse what is happening here. As a paradigm, they considered how flying birds arrive at a collective decision to make a sudden and synchronised change to land. The birds, such as the starlings in the image on the right, have no decision-making leader, yet the flock know exactly how to land in a unified way. The need for the group to land overrides deviating intentions by individual birds. The particle model found that the collective shift to landing depends on perturbations that apply to the individual birds, such as where the birds are in the flock. It is behaviour that can be compared with the way that sand avalanches, if it is piled up, before the point at which symmetric and carefully placed grains would avalanche, because the fluctuations become increasingly non-linear.

"Our main motivation was to better understand something which is puzzling and out there in nature, especially in cases involving the stopping or starting of a collective behavioural pattern in a group of people or animals ... We propose a simple model for a system whose members have the tendency to follow the others both in space and in their state of mind concerning a decision about stopping an activity. This is a very general model, which can be applied to similar situations." The model could also be applied to a swarm of unmanned drones, to initiating a desired motion in a crowd of people, or to interpreting group patterns when stock market shares are bought or sold.

Other Examples

SPP models have been applied in many other areas, such as schooling fish, robotic swarms,molecular motors, the development of human stampedes and the evolution of human trails in urban green spaces.

Swarm (Simulation)

Swarm is the name of a multi-agent simulation package, useful for simulating the interaction ofagents (social or biological) and their emergent collective behaviour. Swarm was initially developed at the Santa Fe Institute in the mid-1990s, and since 1999 has been maintained by the non-profit Swarm Development Group.

Chapter 4

Concepts and Principles

School of Thought

A school of thought is a collection or group of people who share common characteristics of opinionor outlook of a philosophy, discipline, belief, social movement, cultural movement, or art movement.

Schools are often characterized by their currency, and thus classified into "new" and "ol≤ schools. This dichotomy is often a component of paradigm shift. However, it is rarely the case that there are only two schools in any given field. Schools are often named after their founders such as the "Rinzai school" of Zen named after Linji and the Asharite school of early Muslim philosophy named after Abu l'Hasan al-Ashari. They are often also named after their places of origin, such as the Ionian School of philosophy that originated in Ionia and the Chicago school of architecture that originated in Chicago, Illinois and the Prague School of linguistics, named after a linguistic circle found in Prague, or Tartu-Moscow Semiotic School whose representatives lived in Tartu and Moscow.

Mindset

In decision theory and general systems theory, a mindset is a set of assumptions, methods or notations held by one or more people or groups of people which is so established that it creates a powerful incentive within these people or groups to continue to adopt or accept prior behaviors, choices, or tools. This phenomenon of cognitive bias is also sometimes described as *mental inertia*, "groupthink", or a "paradigm", and it is often difficult to counteract its effects upon analysis and decision making processes. On the positive side a mindset can also be seen as incident of a person's Weltanschauung orphilosophy of life. For example there has been quite some interest in the typical mindset of an entrepreneur.

Mindsets in Politics

A well-known example is the "Cold War mindset" prevalent in both the U.S. and USSR, which included absolute trust in two-player game theory, in the integrity of command chain, in control of nuclear materials, and in the mutual assured destruction of both in the case of war. Although most consider that this mindset usefully served to prevent an attack by either country, the assumptions underlyingdeterrence theory have made assessments of the efficacy of the Cold War mindset a matter of some controversy.

Most theorists consider that the key responsibility of an embedded power group is to challenge the assumptions which comprise the group's own mindset. According to these commentators, power groups which fail to review or revise their mindsets with sufficient regularity cannot hold power indefinitely, as a single mindset is unlikely to possess the flexibility and adaptability needed to address all future events. For example, the variations in mindset between Democratic Party andRepublican Party Presidents in the U.S. may have made that country more able to challenge assumptions than the Kremlin with its more static bureaucracy.

Modern military theory attempts to challenge entrenched mindsets in dealing with asymmetric warfare,terrorism and the proliferation of weapons of mass destruction. In combination, these threats represent "a revolution in military affairs" and require very rapid adaptation to new threats and circumstances. In this context, the cost of not implementing adaptive mindsets cannot be afforded.

Collective Mindsets

Naturally, the question regarding the embodiment of a collective mindset comes to mind. Erikson's (1974) analysis of group-identities and what he calls a life-plan seems relevant here. He recounts the example of American Indians, who were meant to undergo a reeducation process meant to imbue a modern "life-plan" which aimed for a house and a richness expressed by filled bank account. Erikson writes that the Indians' collective historic identity as buffalo hunters was oriented around such fundamentally different reasons/goals that even communication about the divergent "life plans" was itself difficult.

There is a double relation between the institution embodying for example an entrepreneurial mindsetand its entrepreneurial performance. Firstly, an institution with an entrepreneurial philosophy will set entrepreneurial goals and strategies as a whole, but maybe even more importantly, it will foster an entrepreneurial milieu, allowing each entity to pursue emergent opportunities. In short, philosophical stance codified in the mind hence as mindset lead to a climate which in turn causes values which lead to practice.

Collective mindsets in this sense are described in such works as Hutchin's "Cognition in the wild" (1995), who analyzes a whole team of naval navigators as the cognitive unit or as computational system, or Senges' Knowledge entrepreneurship in universities (2007). There are also parallels to the emerging field of "collective intelligence" (e.g. (Zara, 2004)) and exploiting the "Wisdom of the crowds" (Surowiecki, 2005) of stakeholders. Zara notes that since collective reflection is more explicit, discursive and conversational it therefore needs a good ¿gestell?—especially when it comes to information and communication technology.

Limit of a Sequence

n	$n \sin(1/n)$
1	0.841471
2	0.958851
...	
10	0.998334
...	
100	0.999983

As the positive integer n becomes larger and larger, the value $n \sin(1/n)$ becomes arbitrarily close to 1. We say that "the limit of the sequence$n$ $\sin(1/n)$ equals 1."

The limit of a sequence (x_n) is, intuitively, the unique number or point L (if it exists) such that the terms of the sequence become arbitrarily close to $L$for "large" values of n. If the limit exists, then we say that the sequence is convergent and that itconverges to L.

Convergence of sequences is a fundamental notion inmathematical analysis, which has been studied since ancient times.

Formal Definition

- For a sequence of real numbers $(x_n : n \in \mathbb{N})$ A real number L is said to be the limit of the sequence x_n, written $\lim_{n\to\infty} x_n = L$, if and only if for every real number $\varepsilon > 0$, there exists a natural number N such that for every natural number $n > N$ we have $| x_n "L | < \varepsilon$.
- As a generalization of this, for a sequence of points $(x_n : n \in \mathbb{N})$ in a topological space T:

An element $L \in T$ is said to be a limit of this sequence if and only if for every neighborhood S of Lthere is a natural number N such that $x_n \in S$ for all $n > N$. In this generality a sequence may admit more than one limit, but if T is a Hausdorff space, then each sequence has at most one limit. When a unique limit L exists, it may be written

$$\lim_{n\to\infty} x_n = L.$$

If a sequence has a limit, we say the sequence is convergent, and that the sequence converges to the limit. Otherwise, the sequence is divergent (see also oscillation).

A null sequence is a sequence that converges to 0.

Hyperreal Definition

A sequence x_n tends to L if for every infinite hypernatural H, the term x_H is infinitely close to L, i.e., the difference x_H - L is infinitesimal. Equivalently, L is the standard part of x_H

$$L = \mathrm{st}(x_H).$$

Thus, the limit can be defined by the formula

$$\lim_{n\to\infty} x_n = \mathrm{st}(x_H),$$

where the limit exists if and only if the righthand side is independent of the choice of an infinite H.

Comments

The definition means that eventually all elements of the sequence get as close as we want to the limit. (The condition that the elements become arbitrarily close to all of the following elements does *not*, in general, imply the sequence has a limit. See Cauchy sequence).

A sequence of real numbers may tend to $+\infty$ or $-\infty$, compare infinite limits. Even though this can be written in the form

$\lim_{n\to\infty} x_n = \infty$ and $\lim_{n\to\infty} x_n = -\infty$

such a sequence is called divergent, unless we explicitly consider it a sequence in the affinely extended real number system or (in the first case only) the real projective line. In the latter cases the sequence has a limit (in the space itself), so could be called convergent, but when using this term here, care should be taken that this does not cause confusion.

The limit of a sequence of points $(x_n : n \in \mathbb{N})$ in a topological space T is a special case of thelimit of a function: the domain is $\mathbb{N}$ in the space $\mathbb{N} \cup \{+\infty\}$ with the induced topology of theaffinely extended real number system, the range is T, and the function argument n tends to +", which in this space is a limit point of .

Examples

- The sequence 1, -1, 1, -1, 1, ... is oscillatory.
- The series with partial sums 1/2, 1/2 + 1/4, 1/2 + 1/4 + 1/8, 1/ 2 + 1/4 + 1/8 + 1/16, ... converges with limit 1. This is an example of an infinite series.
- If a is a real number with absolute value $| a | < 1$, then the sequence a^n has limit 0. If $0 < a$, then the sequence $a^{1/n}$ has limit 1.

Also:

$$\lim_{n\to\infty} \frac{1}{n^p} = 0 \text{ if } p > 0$$

$$\lim_{n\to\infty} a^n = 0 \text{ if } |a| < 1$$

$$\lim_{n\to\infty} n^{\frac{1}{n}} = 1$$

$$\lim_{n\to\infty} a^{\frac{1}{n}} = 1 \text{ if } a > 0$$

Properties

Consider the following function: $f(x) = x_n$ if $n\text{-}1 < x \leq n$. Then the limit of the sequence of x_n is just thelimit of $f(x)$ at infinity.

A function f, defined on a first-countable space, is continuous if and only if it is compatible with limits in that $(f(x_n))$ converges to $f(L)$ given that (x_n) converges to L, *i.e.* $\lim_{n\to\infty} x_n = L$

implies $\lim_{n\to\infty} f(x_n) = f(L)$

Note that this equivalence does *not* hold in general for spaces which are not first-countable.

Compare the basic property (or definition):

f is continuous at x if and only if $\lim_{x\to L} f(x) = f(L)$

A *subsequence* of the sequence (x_n) is a sequence of the form $(x_{a(n)})$ where the $a(n)$ are natural numbers with $a(n) < a(n+1)$ for all n. Intuitively, a subsequence omits some elements of the original sequence. A sequence is convergent if and only if all of its subsequences converge towards the same limit.

Every convergent sequence in a metric space is a Cauchy sequence and hence bounded. A boundedmonotonic sequence of real numbers is necessarily convergent: this is sometimes called the fundamental theorem of analysis.

More generally, every Cauchy sequence of real numbers has a limit, or short: the real numbers are complete.

A sequence of real numbers is convergent if and only if its limit superior and limit inferior coincide and are both finite.

The algebraic operations are continuous everywhere (except for division around zero divisor); thus, given

$$\lim_{n\to\infty} x_n = L_1 \text{ and } \lim_{n\to\infty} y_n = L_2$$

then

$$\lim_{n\to\infty}(x_n + y_n) = L_1 + L_2$$

$$\lim_{n\to\infty}(x_n y_n) = L_1 L_2$$

and (if L_2 and y_n are non-zero)

$$\lim_{n\to\infty}(x_n / y_n) = L_1 / L_2$$

These rules are also valid for infinite limits using the rules

- q + – = – for q `” -”
- q × – = – if $q > 0$
- q × – = -” if $q < 0$
- q / – = 0 if q `” ± –

(see extended real number line).

History

The Greek philosopher Zeno of Elea is famous for formulating paradoxes that involve limiting processes.

Leucippus, Democritus, Antiphon, Eudoxus and Archimedes developed the method of exhaustion, which uses an infinite sequence of approximations to determine an area or a volume. Archimedes succeeded in summing what is now called a geometric series.

Newton dealt with series in his works on *Analysis with infinite series* (written in 1669, circulated in manuscript, published in 1711), *Method of fluxions and infinite series* (written in 1671, published in English translation in 1736, Latin original published much later) and *Tractatus de Quadratura Curvarum*(written in 1693, published in 1704 as an Appendix to his *Optiks*). In the latter work, Newton considers the binomial expansion of $(x+o)^n$ which he then linearizes by *taking limits* (letting $o\to 0$).

In the 18th century, mathematicians like Euler succeeded in summing some *divergent* series by stopping at the right moment; they did not much care whether a limit existed, as long as it could be calculated. At the end of the century, Lagrange in his *Théorie des fonctions analytiques* (1797) opined that the lack of rigour precluded further development in calculus. Gauss in his etude of hypergeometric series (1813) for the first time rigorously investigated under which conditions a series converged to a limit.

The modern definition of a limit (for any ε there exists an index N so that ...) was given independently by Bernhard Bolzano (*Der binomische Lehrsatz*, Prague 1816, little noticed at the time) and byCauchy in his *Cours d'analyse* (1821).

Modes of Convergence

In mathematics, there are many senses in which a sequence or a series is said to be convergent. This article describes various modes (senses or species) of convergence in the settings where they are defined. For a list of modes of convergence, see Modes of convergence (annotated index)

Note that each of the following objects is a special case of the types preceding it: sets, topological spaces, uniform spaces, TAGs (topological abelian groups), normed spaces, Euclidean spaces, and the real/complex numbers. Also, note that any metric space is a uniform space.

Elements of a Topological Space

Convergence can be defined in terms of sequences in first-countable spaces. Nets are a generalization of sequences that is useful in spaces which are not first countable. Filters further generalize the concept of convergence.

In metric spaces, one can define Cauchy sequences. Cauchy nets and filters are generalizations touniform spaces. Even more generally, Cauchy spaces are spaces in which Cauchy filters may be defined. Convergence implies "Cauchy-convergence", and Cauchy-convergence, together with the existence of a convergent subsequence implies convergence. The concept of completeness of metric spaces, and its generalizations is defined in terms of Cauchy sequences.

Series of Elements in a Topological Abelian Group

In a topological abelian group, convergence of a series is defined as convergence of the sequence of partial sums. An important concept when considering series is unconditional convergence, which guarantees that the limit of the series is invariant under permutations of the summands.

In a normed vector space, one can define absolute convergence as convergence of the series of norms ($\Sigma \mid b_k \mid$). Absolute convergence implies Cauchy convergence of the sequence of partial sums (by the triangle inequality), which in turn implies absolute-convergence of some grouping (not reordering). The sequence of partial sums obtained by grouping is a subsequence of the partial sums of the original series. The norm convergence of absolutely convergent series is an equivalent condition for a normed linear space to be Banach (i.e.: complete).

Absolute convergence and convergence together imply unconditional convergence, but unconditional convergence does not imply absolute convergence in general, even if the space is Banach, although the implication holds in $\mathbb{R}^d$.

Convergence of Sequence of Functions to a Topological Space

The most basic type of convergence for a sequence of functions (in particular, it does not assume any topological structure on the domain of the functions) is pointwise convergence. It is defined as convergence of the sequence of values of the functions at every point. If the functions take their values in a uniform space, then one can define pointwise Cauchy convergence, uniform convergence, anduniform Cauchy convergence of the sequence.

Pointwise convergence implies pointwise Cauchy-convergence, and the converse holds if the space in which the functions take their values is complete. Uniform convergence implies pointwise convergence and uniform Cauchy convergence. Uniform Cauchy convergence and pointwise convergence of a subsequence imply uniform convergence of the sequence, and if the codomain is complete, then uniform Cauchy convergence implies uniform convergence.

If the domain of the functions is a topological space, local uniform convergence (i.e. uniform convergence on a neighborhood of each point) and compact (uniform) convergence (i.e. uniform convergence on all compact subsets) may be defined. Note that "compact convergence" is always short for "compact uniform convergence," since "compact pointwise convergence" would mean the same thing as "pointwise convergence" (points are always compact).

Uniform convergence implies both local uniform convergence and compact convergence, since both are local notions while uniform convergence is global. If X is locally compact (even in the weakest sense: every point has compact neighborhood), then local uniform convergence is equivalent to compact (uniform) convergence. Roughly speaking, this is because "local" and "compact" connote the same thing.

Series of Functions to a Topological Abelian Group

Pointwise and uniform convergence of series of functions are defined in terms of convergence of the sequence of partial sums.

For functions taking values in a normed linear space, absolute convergence refers to convergence of the series of positive, real-valued functions $\Sigma \mid g_k \mid$. "Pointwise absolute convergence" is then simply pointwise convergence of $\Sigma \mid g_k \mid$.

Normal convergence is convergence of the series of non-negative real numbers obtained by taking the uniform (i.e. "sup") norm of each function in the series (uniform convergence of $\Sigma \mid g_k \mid$). InBanach spaces, pointwise absolute convergence implies pointwise convergence, and normal convergence implies uniform convergence.

For functions defined on a topological space, one can define (as above) local uniform convergence andcompact (uniform) convergence in terms of the partial sums of the series. If, in addition, the functions take values in a normed linear space, then local normal convergence (local, uniform, absolute convergence) and compact normal convergence (absolute convergence on compact sets). Normal convergence implies both local normal convergence and compact normal convergence. And if the domain is locally compact (even in the weakest sense), then local normal convergence implies compact normal convergence.

Functions Defined on a Measure Space

If one considers sequences of measurable functions, then several modes of convergence that depend on measure-theoretic, rather than solely topological properties, arise. This includes pointwise convergence almost-everywhere, convergence in p-mean and convergence in measure. These are of particular interest in probability theory.

Premature convergence

In genetic algorithms, the term of premature convergence means that a population for anoptimization problem converged too early, resulting in being suboptimal. In this context, the parental solutions, through the aid of genetic operators, are not able to generate offsprings that are superior to their parents. Premature convergence can happen in case of loss of genetic variation (every individual in the population is identical, see convergence).

Strategies for preventing premature convergence

Strategies to regain genetic variation can be:

- a mating strategy called *incest prevention*,
- uniform crossover,
- favored replacement of similar individuals (*preselection* or *crowding*),
- segmentation of individuals of similar fitness (*fitness sharing*),
- increasing population size.

The genetic variation can also be regained by mutation though this process is highly random.

Local Optimum

Local optimum is a term in applied mathematics andcomputer science.

A local optimum of a combinatorial optimization problem is a solution that is optimal (either maximal or minimal) within a neighboring set of solutions. This is in contrast to a global optimum, which is the optimal solution among all possible solutions.

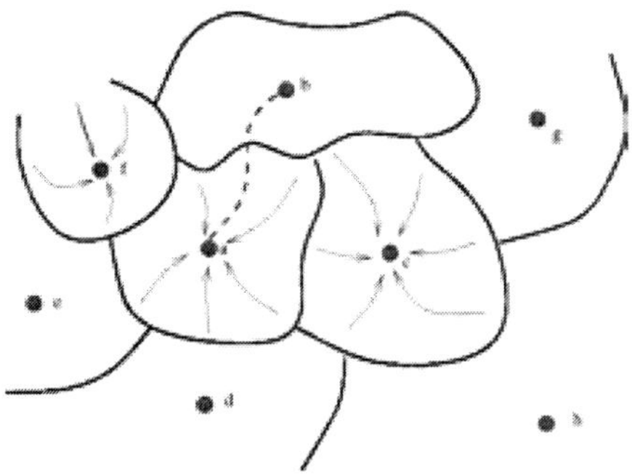

Figure: Attraction basins around locally optimal points

Local search or hill climbing methods for solving discrete optimization problems start from an initial configuration and repeatedly move to an *improving neighboring configuration*. A trajectory in search space is generated, which maps an initial point to a local optimum, where local search is stuck (no improving neighbors are available). The search space is therefore subdivided into *attraction basins*, consisting of all initial points which have a given local optimum as final point of the local search trajectory. A local optimum can be isolated (surrounded by non locally-optimal point) or part of aplateau, a locally optimal region with more than one point.

If the problem to be solved has all local optimal points with the same value of the function to be optimized, local search effectively solves the problem: finding a local optimum delivers the globally optimal solution.

The locality of the optimum is dependent on the neighborhood structure as defined by the local searchmethod that is used for optimizing the solution.

In many cases, local optima deliver sub-optimal solutions, and a local search method needs to be modified to continue the search beyond local optimality, see for example iterated local search, tabu search, reactive search optimization, simulated annealing.

Convergence (Evolutionary Computing)

Precisely every individual in the population is identical. While full convergence might be seen ingenetic algorithms using only cross over, such *convergence* is seldom seen in genetic programmingusing Koza's subtree swapping crossover. However, populations often stabilise after a time, in the sense that the best programs all have a common ancestor and their behaviour is very similar (or identical) both to each other and to that of high fitness programs from the previous (and future?) generations. Often the term *convergence* is loosely used.

Multi-swarm Optimization

Multi-swarm optimization is a variant of particle swarm optimization (PSO) based on the use of multiple sub-swarms instead of one (standard) swarm. The general approach in multi-swarm optimization is that each sub-swarm focuses on a specific region while a specific diversification method decides where and when to launch the sub-swarms. The multi-swarm framework is especially fitted for the optimization on multi-modal problems, where multiple (local) optima exist.

Description

In multi-modal problems it is important to achieve an effective balance between exploration and exploitation. Multi-swarm systems provide a new approach to improve this balance. Instead of trying to achieve a compromise between exploration and exploitation which could weaken both mechanisms of the search process, multi-swarm systems separate them into distinct phases. Each phase is more focused on either exploitation (individual sub-swarms) or exploration (diversification method).

The coordination of the sub-swarms depends on the specific diversification method(s) implemented by the multi-swarm system. Wave of Swarm of Particles (WOSP) , for example, bases its diversification mechanism on the "collision" of particles. When particles get too close they are expeled by a short range force into new waves/ sub-swarms, avoiding thus a complete convergence. The Dynamic Multi-Swarm-Particle Swarm Optimizer (DMS-PSO) periodically regroups the particles of the sub-swarms (after they have converged) into new sub-swarms, the new swarms are started with particles from previous swarms. Locust swarms are based on a "devour and move on" strategy – after a sub-swarm "devours" a relatively small region of the search space (to find a local optimum) scouts are deployed to look for new promising regions to "move on".

A distinctive feature of sub-swarms is that their initial positions and initial velocities are not randomly selected as in normal swarms. Instead, they maintain some information from the previous trajectories of the particles. In general, the development of multi-swarm systems leads to design decisions which did not exist during the original development of particle swarm optimization, such as the number of particles to use in each sub-swarm, the optimal value for the constriction factor and the effects of non-random initial positions and initial velocities. These design decisions have been thoroughly studied and have well-established guidelines – e.g. the use of non-random initial positions and initial velocities leads to improved results in multi-swarm systems, which is not the case for single-swarms . Other design decisions, such as which diversification method to use or which specific search strategy will select the initial positions and initial velocities of a sub-swarm, have less established guidelines and constitute open questions in the field of multi-swarm systems.

Occam's Razor

Occam's razor, also known as Ockham's razor, and sometimes expressed in Latin as *lex parsimoniae*(the law of parsimony, economy or succinctness), is a principle that generally recommends from among competing hypotheses selecting the one that makes the fewest new assumptions.

Overview

The principle is often summarized as "simpler explanations are, other things being equal, generally better than more complex ones." In practice, the principle is usually focused on shifting the burden of proof in discussions. That is, the razor is a principle that suggests we should *tend* towards simpler theories until we can trade some simplicity for increased explanatory power. Contrary to the popular summary, the simplest available theory is sometimes a less accurate explanation. Philosophers also add that the exact meaning of "simplest" can be nuanced in the first place.

Bertrand Russell offered what he called "a form of Occam's Razor": "Whenever possible, substitute constructions out of known entities for inferences to unknown entities."

Occam's razor is attributed to the 14th-century English logician, theologian and Franciscan friar FatherWilliam of Ockham (d'Okham) although the principle was familiar long before. The words attributed to Occam are "entities must not be multiplied beyond necessity" (*entia non sunt multiplicanda praeter necessitatem*), although these actual words are not to be found in his extant works. The saying is also phrased as *pluralitas non est ponenda sine necessitate* ("plurality should not be posited without necessity"). To quote Isaac Newton, "We are to admit no more causes of natural things than such as are both true and sufficient to explain their appearances. Therefore, to the same natural effects we must, so far as possible, assign the same causes."

In science, Occam's razor is used as a heuristic (general guiding rule or an observation) to guide scientists in the development of theoretical models rather than as an arbiter between published models. In the scientific method, Occam's razor is not considered an irrefutable principle of logic, and certainly not a scientific result.

Solomonoff's inductive inference is a mathematical proof of Occam's razor, under the assumption that the environment follows some unknown but computable probability distribution.

History

William of Ockham (*c.* 1285–1349) is remembered as an influential nominalist though his popular fame as a great logician rests chiefly on the maxim attributed to him and known as Ockham's razor. The term *razor* (the German "Ockhams Messer" translates to "Occam's knife") refers to distinguishing between two theories either by "shaving away" unnecessary assumptions or cutting apart two similar theories.

This maxim seems to represent the general tendency of Occam's philosophy, but it has not been found in any of his writings. His nearest pronouncement seems to be *Numquam ponenda est pluralitas sine necessitate* [Plurality must never be posited without necessity], which occurs in his theological work on the *Sentences of Peter Lombard* (*Quaestiones et decisiones in quattuor libros Sententiarum Petri Lombardi* (ed. Lugd., 1495), i, dist. 27, qu. 2, K).

In his *Summa Totius Logicae*, i. 12, Ockham cites the principle of economy, *Frustra fit per plura quod potest fieri per pauciora* [It is futile to do with more things that which can be done with fewer].

The origins of what has come to be known as Occam's razor are traceable to the works of earlier philosophers such asMaimonides (Rabbi Moshe ben Maimon, 1138–1204), John Duns Scotus (1265–1308), and even Aristotle (384–322 BC) (Charlesworth 1956).

The term "Occam's razor" first appeared in 1852 in the works of Sir William Hamilton, 9th Baronet (1788–1856), centuries after Ockham's death. Ockham did not invent this "razor"; its association with him may be due to the frequency and effectiveness with which he used it (Ariew 1976). Ockham stated the principle in various ways, but the most popular version was written by John Ponce from Cork in 1639 (Meyer 1957).

For Ockham, the only truly necessary entity is God; everything else, the whole of creation, is radically contingent through and through.

Justifications

Beginning in the 20th century, epistemological justifications based on induction, logic, pragmatism, and especially probability theory have become more popular among philosophers.

Aesthetic

Prior to the 20th century, it was a commonly-held belief that nature itself was simple and that simpler hypotheses about nature were thus more likely to be true. This notion was deeply rooted in the aesthetic value simplicity holds for human thought and the justifications presented for it often drew from theology. Thomas Aquinas made this argument in the 13th century, writing, "If a thing can be done adequately by means of one, it is superfluous to do it by means of several; for we observe that nature does not employ two instruments [if] one suffices."

Linguistic

Simon argued that whether something is simple or complex depends on the way we choose to describe it. Quine proposed his Maxim of Shallow Analysis which says that we should uncover no more structure than necessary in order to show that a sentence is grammatical.

Empirical

Occam's razor has gained strong empirical support as far as helping to converge on better theories (see "Applications" section below for some examples).

Even if Occam's razor is empirically justified, so too is the need to use other "theory selecting" methods in science. Such other scientific methods are what support the razor's validity as a tool in the first place. This is because measuring the razor's (or any method's) ability to select between theories requires the use of different, reliable "theory selecting" methods for corroboration.

One should note the related concept of overfitting, where excessively complex models are affected bystatistical noise (a problem also known as the bias-variance trade-off), whereas simpler models may capture the underlying structure better and may thus have better predictive performance. It is, however, often difficult to deduce which part of the data is noise (cf. model selection, test set, minimum description length, Bayesian inference, etc.).

Testing the Razor

The razor's claim that "simpler explanations are, other things being equal, generally better than more complex ones" is amenable to empirical testing. The procedure to test this hypothesis would compare the track records of simple and comparatively complex explanations. The validity of Occam's razor as a tool would then have to be rejected if the more complex explanations were more often correct than the less complex ones (while the converse would lend support to its use).

Figure: Possible explanations can get needlessly complex. It is coherent, for instance, to add the involvement ofLeprechauns to any explanation, but Occam's razor would prevent such additions, unless they were necessary.

In the history of competing explanations this is certainly not the case. At least, not generally (some increases in complexity are sometimes necessary), and so there remains a justified general bias towards the simpler of two competing explanations. To understand why, consider that, for each accepted explanation of a phenomenon, there is always an*infinite* number of possible, more complex, and ultimately incorrect alternatives. This is so because one can always burden failing explanations with ad-hoc hypotheses. Ad-hoc hypotheses are justifications which prevent theories from beingfalsified. Even other empirical criteria like consilience can never truly eliminate such explanations as competition. Each true explanation, then, may have had *many* alternatives that were simpler and false, but also an *infinite* number of alternatives that were more complex and false. Put another way, any new, and even more complex theory can still possibly be true. For example: If an individual makessupernatural claims that Leprechauns were responsible for breaking a vase, the simpler explanation would be that he is mistaken, but ongoing ad-hoc justifications (e.g. "And, that's not me on film, they tampered with that too") successfully prevent outright falsification. This endless supply of elaborate competing explanations cannot be ruled out—but by using Occam's Razor.

Practical Considerations and Pragmatism

The common form of the razor, used to distinguish between equally explanatory hypotheses, may be supported by the practical fact that simpler theories are easier to understand.

Some argue that Occam's razor is not a theory at all (in the classic sense of being an inference-driven model); rather, it may be a heuristic maxim for choosing *among* other theories and instead *underlies*induction.

Alternatively, if we want to have reasonable discussion we may be practically forced to accept Occam's razor in the same way we are simply forced to accept the laws of thought and inductive reasoning (given the problem of induction). As philosopher Elliott Sober explains (see below) not even Reason itself can be justified on any reasonable grounds. This has been taken to prove that the accepted bedrock premises of understanding are *necessarily* unjustifiable by pure reason; we must start with first principles of some kind (otherwise an infinite regress occurs).

The pragmatist may go on, as David Hume did on the topic induction, that there is no satisfying alternative to granting this premise. Though one may *claim* that Occam's razor is invalid as a premise helping to regulate theories, putting this doubt into practice would mean doubting whether every step forward will result in locomotion or a nuclear explosion. In other words still: "What's the alternative?"

Mathematical

There have been attempts to derive Occam's Razor from probability theory, notable attempts made byHarold Jeffreys and E. T. Jaynes. Using Bayesian reasoning, a simple theory is preferred to a complicated one because of a higher prior probability. William H. Jeffreys and

Berger stated that "as a consequence of the fact that a hypothesis with fewer adjustable parameters will automatically have an enhanced posterior probability, due to the fact that the predictions it makes are sharp..."

Other Views

Karl Popper

Karl Popper argues that a preference for simple theories need not appeal to practical or aesthetic considerations. Our preference for simplicity may be justified by its falsifiability criterion: We prefer simpler theories to more complex ones "because their empirical content is greater; and because they are better testable" (Popper 1992). The idea here is that a simple theory applies to more cases than a more complex one, and is thus more easily falsifiable. This is again comparing a simple theory to a more complex theory where both explain the data equally well.

Elliott Sober

The philosopher of science Elliott Sober once argued along the same lines as Popper, tying simplicity with "informativeness": The simplest theory is the more informative one, in the sense that less information is required in order to answer one's questions (Sober 1975). He has since rejected this account of simplicity, purportedly because it fails to provide an epistemic justification for simplicity. He now expresses views to the effect that simplicity considerations (and considerations of parsimony in particular) do not count unless they reflect something more fundamental. Philosophers, he suggests, may have made the error of hypostatizing simplicity (i.e. endowed it with a *sui generis* existence), when it has meaning only when embedded in a specific context (Sober 1992). If we fail to justify simplicity considerations on the basis of the context in which we make use of them, we may have no non-circular justification: "just as the question 'why be rational?' may have no non-circular answer, the same may be true of the question 'why should simplicity be considered in evaluating the plausibility of hypotheses?'" (Sober 2001)

Richard Swinburne

Richard Swinburne argues for simplicity on logical grounds:

... the simplest hypothesis proposed as an explanation of phenomena is more likely to be the true one than is any other available hypothesis, that its predictions are more likely to be true than those of any other available hypothesis, and that it is an ultimate a priori epistemic principle that simplicity is evidence for truth.

—Swinburne 1997

Since our choice of theory cannot be determined by data (see Underdetermination and Quine-Duhem thesis), we must rely on some criterion to determine which theory to use. Since it is absurd to have no logical method by which to settle on one hypothesis amongst an infinite number of equally data-compliant hypotheses, we should choose the simplest theory: "...either science is irrational [in the way it judges theories and predictions probable] or the principle of simplicity is a fundamental synthetic a priori truth" (Swinburne 1997).

Ludwig Wittgenstein

- 3.328 If a sign is not necessary then it is meaningless. That is the meaning of Occam's razor. (If everything in the symbolism works as though a sign had meaning, then it has meaning.)
- 4.04 In the proposition there must be exactly as many things distinguishable as there are in the state of affairs which it represents. They must both possess the same logical (mathematical) multiplicity (cf. Hertz's Mechanics, on Dynamic Models).
- 5.47321 Occam's razor is, of course, not an arbitrary rule nor one justified by its practical success. It simply says that unnecessary elements in a symbolism mean nothing. Signs which serve one purpose are logically equivalent, signs which serve no purpose are logically meaningless. and on the related concept of "simplicity":
- 6.363 The procedure of induction consists in accepting as true the simplest law that can be reconciled with our experiences.

Applications

Science and the Scientific Method

In science, Occam's razor is used as a heuristic (rule of thumb) to guide scientists in the development of theoretical models rather than as an arbiter between published models. In physics, parsimonywas an important heuristic in the formulation of special relativity by Albert Einstein, the development and application of the principle of least action by Pierre Louis Maupertuis and Leonhard Euler, and the development of quantum mechanics by Ludwig Boltzmann, Max Planck, Werner Heisenberg and Louis de Broglie. In chemistry, Occam's razor is often an important heuristicwhen developing a model of a reaction mechanism. However, while it is useful as a heuristic in developing models of reaction mechanisms, it has been shown to fail as a criterion for selecting among some selected published models. In this context, Einstein himself expressed a certain caution when he formulated Einstein's Constraint: "Everything should be kept as simple as possible, but no simpler." Elsewhere, Einstein harks back to the theological roots of the razor, with his famous put-down: "The Good Lord may be subtle, but he is not malicious."

In the scientific method, parsimony is an epistemological, metaphysical or heuristic preference, not an irrefutable principle of logic, and certainly not a scientific result. As a logical principle, Occam's razor would demand that scientists accept the simplest possible theoretical explanation for existing data. However, science has shown repeatedly that future data often supports more complex theories than existing data. Science tends to prefer the simplest explanation that is consistent with the data available at a given time, but history shows that these simplest explanations often yield to complexities as new data become available. Science is open to the possibility that future experiments might support more complex theories than demanded by current data and is more interested in designing experiments to discriminate between competing theories than favoring one theory over another based merely on philosophical principles.

When scientists use the idea of parsimony, it only has meaning in a very specific context of inquiry. A number of background assumptions are required for parsimony to connect with plausibility in a particular research problem. The reasonableness of parsimony in one research context may have nothing to do with its reasonableness in another. It is a mistake to think that there is a single global principle that spans diverse subject matter.

As a methodological principle, the demand for simplicity suggested by Occam's razor cannot be generally sustained. Occam's razor cannot help toward a rational decision between competing explanations of the same empirical facts. One problem in formulating an explicit general principle is that complexity and simplicity are perspective notions whose meaning depends on the context of application and the user's prior understanding. In the absence of an objective criterion for simplicity and complexity, Occam's razor itself does not support an objective epistemology.

The problem of deciding between competing explanations for empirical facts cannot be solved by formal tools. Simplicity principles can be useful heuristics in formulating hypotheses, but they do not make a contribution to the selection of theories. A theory that is compatible with one person's world view will be considered simple, clear, logical, and evident, whereas what is contrary to that world view will quickly be rejected as an overly complex explanation with senseless additional hypotheses. Occam's razor, in this way, becomes a "mirror of prejudice."

It has been suggested that Occam's razor is a widely accepted example of extraevidential consideration, even though it is entirely a metaphysical assumption. There is little empirical evidence that the world is actually simple or that simple accounts are more likely than complex ones to be true.

Most of the time, Occam's razor is a conservative tool, cutting out crazy, complicated constructions and assuring that hypotheses are grounded in the science of the day, thus yielding 'normal' science: models of explanation and prediction. There are, however, notable exceptions where Occam's razor turns a conservative scientist into a reluctant revolutionary. For example, Max Planck interpolated between the Wien and Jeans radiation laws used an Occam's razor logic to formulate the quantum hypothesis, and even resisting that hypothesis as it became more obvious that it was correct.

However, on many occasions Occam's razor has stifled or delayed scientific progress. For example, appeals to simplicity were used to deny the phenomena of meteorites, ball lightning,continental drift, and reverse transcriptase. It originally rejected DNA as the carrier of genetic information in favor of proteins, since proteins provided the simpler explanation. Theories that reach far beyond the available data are rare, but general relativity provides one example.

In hindsight, one can argue that it is simpler to consider DNA as the carrier of genetic information, because it uses a smaller number of building blocks (four nitrogenous bases). However, during the time that proteins were the favored genetic medium, it seemed like a more complex hypothesis to confer genetic information in DNA rather than proteins. One can also argue (also in hindsight) for atomic building blocks for matter, because it provides a simpler explanation for the observed reversibility of both mixing and chemical reactions as simple separation and re-arrangements of the atomic building blocks. However, at the time, the atomic theorywas considered more complex because it inferred the existence of invisible particles which had not been directly detected. Ernst Mach and the logical positivists rejected the atomic theory of John Dalton, until the reality of atoms was more evident in Brownian motion, as explained by Albert Einstein.

In the same way, hindsight argues that postulating the aether is more complex than transmission of light through a vacuum. However, at the time, all known waves propagated through a physical medium, and it seemed simpler to postulate the existence of a medium rather than theorize about wave propagation without a medium. Likewise, Newton's idea of light particles seemed simpler than Christiaan Huygens's idea of waves, so many favored it; however in this case, as it turned out, neither the wave- nor the particle-explanation alone suffices, since light behaves like waves as well as like particles (wave–particle duality).

Three axioms presupposed by the scientific method are realism (the existence of objective reality), the existence of natural laws, and the constancy of natural law. Rather than depend on provability of these axioms, science depends on the fact that they have not been objectively falsified. Occam's razor andparsimony support, but do not prove these general axioms of science. The general principle of science is that theories (or models) of natural law must be consistent with repeatable experimental observations. This ultimate arbiter (selection criterion) rests upon the axioms mentioned above.

There are many examples where Occam's razor would have picked the wrong theory given the available data. Simplicity principles are useful philosophical preferences for choosing a more likely theory from among several possibilities that are each consistent with available data. A single instance of Occam's razor picking a wrong theory falsifies the razor as a general principle.

If multiple models of natural law make exactly the same testable predictions, they are equivalent and there is no need for parsimony to choose one that is preferred. For example, Newtonian, Hamiltonian, and Lagrangian classical mechanics are equivalent. Physicists have no interest in using Occam's razor to say the other two are wrong. Likewise, there is no demand for simplicity principles to arbitrate between wave and matrix formulations of quantum mechanics. Science often does not demand arbitration or selection criteria between models which make the same testable predictions.

Michael Lee and others provide cases where a parsimonious approach does not guarantee a correct conclusion and, if based on incorrect working hypotheses or interpretations of incomplete data,*may even strongly support a false conclusion.* He warns "When parsimony ceases to be a guideline and is instead elevated to an *ex cathedra* pronouncement, parsimony analysis ceases to be science."

Biology

Biologists or philosophers of biology use Occam's razor in either of two contexts both in evolutionary biology: the units of selection controversy and systematics. George C. Williams in his book*Adaptation and Natural Selection* (1966) argues that the best way to explain altruism among animals is based on low level (i.e. individual) selection as opposed to high level group selection. Altruism is defined as behavior that is beneficial to the group but not to the individual, and group selection is thought by some to be the evolutionary mechanism that selects for altruistic traits. Others posit individual selection as the mechanism which explains altruism solely in terms of the behaviors of individual organisms acting in their own self interest without regard to the group. The basis for Williams's contention is that of the two, individual selection is the more parsimonious theory. In doing so he is invoking a variant of Occam's razor known as Lloyd Morgan's Canon: "In no case is an animal activity to be interpreted in terms of higher psychological processes, if it can be fairly interpreted in terms of processes which stand lower in the scale of psychological evolution and development" (Morgan 1903).

However, more recent biological analyses, such as Richard Dawkins's *The Selfish Gene*, have contended that Williams's view is not the simplest and most basic. Dawkins argues the way evolution works is that the genes that are propagated in most copies will end up determining the development of that particular species, i.e., natural selection turns out to select specific genes, and this is really the fundamental underlying principle, that automatically gives individual and group selection as emergentfeatures of evolution.

Zoology provides an example. Muskoxen, when threatened by wolves, will form a circle with the males on the outside and the females and young on the inside. This as an example of a behavior by the males that seems to be altruistic. The behavior is disadvantageous to them individually but beneficial to the group as a whole and was thus seen by some to support the group selection theory.

However, a much better explanation immediately offers itself once one considers that natural selection works on genes. If the male musk ox runs off, leaving his offspring to the wolves, his genes will not be propagated. If however he takes up the fight his genes will live on in his offspring. And thus the "stay-and-fight" gene prevails. This is an example of kin selection. An underlying general principle thus offers a much simpler explanation, without retreating to special principles as group selection.

Systematics is the branch of biology that attempts to establish genealogical relationships among organisms. It is also concerned with their classification. There are three primary camps in systematics; cladists, pheneticists, and evolutionary taxonomists. The cladists hold that genealogyalone should determine classification and pheneticists contend that similarity over propinquity of descent is the determining criterion while evolutionary taxonomists claim that both genealogy and similarity count in classification.

It is among the cladists that Occam's razor is to be found, although their term for it is cladistic parsimony. Cladistic parsimony (or maximum parsimony) is a method of phylogenetic inference in the construction of types of phylogenetic trees (more specifically, cladograms). Cladograms are branching, tree-like structures used to represent lines of descent based on one or more evolutionary change (s). Cladistic parsimony is used to support the hypothesis (es) that require the fewest evolutionary changes. For some types of tree, it will consistently produce the wrong results regardless of how much data is collected (this is called long branch attraction). For a full treatment of cladistic parsimony, see Elliott Sober's *Reconstructing the Past: Parsimony, Evolution, and Inference* (1988). For a discussion of both uses of Occam's razor in Biology see Elliott Sober's article *Let's Razor Ockham's Razor* (1990).

Other methods for inferring evolutionary relationships use parsimony in a more traditional way.Likelihood methods for phylogeny use parsimony as they do for all likelihood tests, with hypotheses requiring few differing parameters (i.e., numbers of different rates of character change or different frequencies of character state transitions) being treated as null hypotheses relative to hypotheses requiring many differing parameters. Thus, complex hypotheses must predict data much better than do simple hypotheses before researchers reject the simple hypotheses. Recent advances employinformation theory, a close cousin of likelihood, which uses Occam's razor in the same way.

Francis Crick has commented on potential limitations of Occam's razor in biology. He advances the argument that because biological systems are the products of (an on-going) natural selection, the mechanisms are not necessarily optimal in an obvious sense. He cautions: "*While Ockham's razor is a useful tool in the physical sciences, it can be a very dangerous implement in biology. It is thus very rash to use simplicity and elegance as a guide in biological research.*"

In biogeography, parsimony is used to infer ancient migrations of species or populations by observing the geographic distribution and relationships of existing organisms. Given the phylogenetic tree, ancestral migrations are inferred to be those that require the minimum amount of total movement.

Medicine

When discussing Occam's razor in contemporary medicine, doctors and philosophers of medicine speak of diagnostic parsimony. Diagnostic parsimony advocates that when diagnosing a given injury, ailment, illness, or disease a doctor should strive to look for the fewest possible causes that will account for all the symptoms. This philosophy is one of several demonstrated in the popular medicaladage "when you are in Texas and you hear hoofbeats, think horses, not zebras." While diagnostic parsimony might often be beneficial, credence should also be given to the counter-argument modernly known as Hickam's dictum, which succinctly states that "patients can have as many diseases as they damn well please." It is often statistically more likely that a patient has several common diseases, rather than having a single rarer disease which explains their myriad symptoms. Also, independently of statistical likelihood, some patients do in fact turn out to have multiple diseases, which by common sense nullifies the approach of insisting to explain any given collection of symptoms with one disease. These misgivings emerge from simple probability theory—which is already taken into account in many modern variations of the razor—and from the fact that the loss function is much greater in medicine than in most of general science. Because misdiagnosis can result in the loss of a person's health and potentially life, it is considered better to test and pursue all reasonable theories even if there is some theory that appears the most likely. Diagnostic parsimony and the counter-balance it finds in Hickam's dictum have very important implications in medical practice.

Any set of symptoms could be indicative of a range of possible diseases and disease combinations; though at no point is a diagnosis rejected or accepted just on the basis of one disease appearing more likely than another, the continuous flow of hypothesis formulation, testing and modification benefits greatly from estimates regarding which diseases (or sets of diseases) are relatively more likely to be responsible for a set of symptoms, given the patient's environment, habits, medical history and so on. For example, if a hypothetical patient's immediately apparent symptoms include fatigue and cirrhosis and they test negative for Hepatitis C, their doctor might formulate a working hypothesis that the cirrhosis was caused by their drinking problem, and then seek symptoms and perform tests to formulate and rule out hypotheses as to what has been causing the fatigue; but if the doctor were to further discover that the patient's breath inexplicably smells of garlic and they are suffering from pulmonary edema, they might decide to test for the relatively rare condition of Selenium poisoning.

Religion

In the philosophy of religion, Occam's razor is sometimes applied to the existence of God; if the concept of a God does not help to explain the universe better, then the idea is that atheism should be preferred (Schmitt 2005). Some such arguments are based on the assertion that belief in God requires more complex assumptions to explain the universe than non-belief (e.g. the Ultimate Boeing 747 gambit). On the other hand, there are various arguments in favour of a God which attempt to establish a God as a useful explanation. Philosopher Del Ratzsch suggests that the application of the razor to God may not be so simple, least of all when we are comparing that hypothesis with theories postulating multiple invisible universes.

God as Beside the Razor

Rather than argue for the necessity of God, some theists consider their belief to be based on grounds independent of, or prior to, reason, making Occam's razor irrelevant. This was the stance of Søren Kierkegaard, who viewed belief in God as a leap of faith which sometimes directly opposed reason.This is also the same basic view of Clarkian Presuppositional apologetics, with the exception that Clark never thought the leap of faith was contrary to reason. (See also: Fideism). In a different vein,Alvin Plantinga and others have argued for reformed epistemology, the view that God's existence can properly be assumed as part of a Christian's epistemological structure (see also basic beliefs). Yet another school of thought, Van Tillian presuppositional apologetics, claims that God's existence is thetranscendentally necessary prior condition to the intelligibility of all human experience and thought. In other words, proponents of this view hold that there is no other viable option to ultimately explain any fact of human experience or knowledge, let alone a simpler one.

William of Ockham himself was a theist. He believed in God, and thus in some validity of scripture; he writes that "nothing ought to be posited without a reason given, unless it is self-evident (literally, known through itself) or known by experience or proved by the authority of Sacred Scripture." In Ockham's view, an explanation which does not harmonize with reason, experience or the aforementioned sources cannot be considered valid. However, unlike many theologians of his time, Ockham did not believe God could be logically proven with arguments. In fact, he thought that science actually seemed to eliminate God according to the Razor's criteria. To Ockham, science was a matter of discovery, but theology was a matter of revelation and faith (e.g. some sort of Non-overlapping magisteria). He explains: "only faith gives us access to theological truths. The ways of God are not open to reason, for God has freely chosen to create a world and establish a way of salvation within it apart from any necessary laws that human logic or rationality can uncover."

Philosophy of Mind

Probably the first person to make use of the principle was Ockham himself. He writes "The source of many errors in philosophy is the claim that a distinct signified thing always corresponds to a distinct word in such a way that there are as many distinct entities being signified as there are distinct names or words doing the signifying." (*Summula Philosophiae Naturalis III*, chap. 7, see also *Summa Totus Logicae* Bk I, C.51). We are apt to suppose that a word like "paternity" signifies some "distinct entity", because we suppose that each distinct word signifies a distinct entity. This leads to all sorts of absurdities, such as "a column is to the right by to-the-rightness", "God is creating by creation, is good by goodness, is just by justice, is powerful by power", "an accident inheres by inherence", "a subject is subjected by subjection", "a suitable thing is suitable by suitability", "a chimera is nothing by nothingness", "a blind thing is blind by blindness", "a body is mobile by mobility". We should say instead that a man is a father because he has a son (Summa C.51).

Another application of the principle is to be found in the work of George Berkeley (1685–1753). Berkeley was an idealist who believed that all of reality could be explained in terms of the mind alone. He famously invoked Occam's razor against Idealism's metaphysical competitor, materialism, claiming that matter was not required by his metaphysic and was thus eliminable. One problem with this argument is that the razor is easily turned around on Berkeley's Idealism itself, which is premised on the notion of a supernatural entity constantly projecting ideas into the observer's mind to give the impression of matter. Invoking such a supposition to explain the appearance of matter is far more unnecessary than the supposition that matter itself is real.

In the 20th century Philosophy of Mind, Occam's razor found a champion in J. J. C. Smart, who in his article "Sensations and Brain Processes" (1959) claimed Occam's razor as the basis for his preference of the mind-brain identity theory over mind-body dualism. Dualists claim that there are two kinds of substances in the universe: physical (including the body) and mental, which is nonphysical. In contrast identity theorists claim that everything is physical, including consciousness, and that there is nothing nonphysical.

The basis for the materialist claim is that of the two competing theories, dualism and mind-brain identity, the identity theory is the simpler since it commits to fewer entities. Smart was criticized for his use of the razor and ultimately retracted his advocacy of it in this context.

Paul Churchland (1984) cites Occam's razor as the first line of attack against dualism, but admits that by itself it is inconclusive. The deciding factor for Churchland is the greater explanatory prowess of a materialist position in the Philosophy of Mind as informed by findings in neurobiology.

Dale Jacquette (1994) claims that Occam's razor is the rationale behind eliminativism and reductionism in the philosophy of mind. Eliminativism is the thesis that the ontology of folk psychologyincluding such entities as "pain", "joy", "desire", "fear", etc., are eliminable in favor of an ontology of a completed neuroscience.

Penal Ethics

In penal theory and the philosophy of punishment, parsimony refers specifically to taking care in the distribution of punishment in order to avoid excessive punishment. In the utilitarian approach to the philosophy of punishment, Jeremy Bentham's "parsimony principle" states that any punishment greater than is required to achieve its end is unjust. The concept is related but not identical to the legal concept of proportionality. Parsimony is a key consideration of the modern restorative justice, and is a component of utilitarian approaches to punishment, as well as the prison abolition movement. Bentham believed that true parsimony would require punishment to be individualised to take account of the sensibility of the individual—an individual more sensitive to punishment should be given a proportionately lesser one, since otherwise needless pain would be inflicted. Later utilitarian writers have tended to abandon this idea, in large part due to the impracticality of determining each alleged criminal's relative sensitivity to specific punishments.

Probability Theory and Statistics

One intuitive justification of Occam's razor's admonition against unnecessary hypotheses is a direct result of basic probability theory. By definition, all assumptions introduce possibilities for error; if an assumption does not improve the accuracy of a theory, its only effect is to increase the probability that the overall theory is wrong.

There are various papers in scholarly journals deriving formal versions of Occam's razor from probability theory and applying it in statistical inference, and also of various criteria for penalizing complexity in statistical inference. Recent papers have suggested a connection between Occam's razor andKolmogorov complexity.

One of the problems with the original formulation of the principle is that it only applies to models with the same explanatory power (i.e. prefer the simplest of equally good models). A more general form of Occam's razor can be derived from Bayesian model comparison and Bayes factors, which can be used to compare models that don't fit the data equally well. These methods can sometimes optimally balance the complexity and power of a model. Generally the exact Ockham factor is intractable but approximations such as Akaike Information Criterion, Bayesian Information Criterion, Variational Bayes, False discovery rate and Laplace approximation are used. Many artificial intelligenceresearchers are now employing such techniques.

William H. Jefferys and James O. Berger (1991) generalise and quantify the original formulation's "assumptions" concept as the degree to which a proposition is unnecessarily accommodating to possible observable data. The model they propose balances the precision of a theory's predictions against their sharpness; theories which sharply made their correct predictions are preferred over theories which would have accommodated a wide range of other possible results. This, again, reflects the mathematical relationship between key concepts in Bayesian inference (namely marginal probability, conditional probability and posterior probability).

The statistical view leads to a more rigorous formulation of the razor than previous philosophical discussions. In particular, it shows that "simplicity" must first be defined in some way before the razor may be used, and that this definition will always be subjective.[*why?*] For example, in the Kolmogorov-Chaitin Minimum description length approach, the subject must pick a Turing machine whose operations describe the basic operations believed to represent "simplicity" by the subject. However one could always choose a Turing machine with a simple operation that happened to construct one's entire theory and would hence score highly under the razor. This has led to two opposing views of the objectivity of Occam's razor.

Objective Razor

The minimum instruction set of a Universal Turing machine requires approximately the same length description across different formulations, and is small compared to the Kolmogorov complexity of most practical theories. Marcus Hutter has used this consistency to define a "natural" Turing machine of small size as the proper basis for excluding arbitrarily complex instruction sets in the formulation of razors. Describing the program for the universal program as the "hypothesis", and the representation of the evidence as program data, it has been formally proven under ZF that "the sum of the log universal probability of the model plus the log of the probability of the data given the model should be minimized." Interpreting this as minimising the total length of a two-part message encoding model followed by data given model gives us the Minimum Message Length (MML) principle

One possible conclusion from mixing the concepts of Kolmogorov complexity and Occam's razor is that an ideal data compressor would also be a scientific explanation/formulation generator. Some attempts have been made to re-derive known laws from considerations of simplicity or compressibility.

According to Jürgen Schmidhuber, the appropriate mathematical theory of Occam's razor already exists, namely, Ray Solomonoff's theory of optimal inductive inference and its extensions. See discussions in for the subtle distinctions between the algorithmic probability (ALP) work of Ray Solomonoff and the Minimum Message Length work of Chris Wallace, and see both for such discussions and also (in sec. 4) discussions of MML and Ockham's razor. For a specific example of MML as Ockham's razor in the problem of decision tree induction, see.

In Literature and Writing

Occam's razor has been recommended as a measure of how good the plot of a novel is. Simple and logical plots are easy to explain and this enhances the experience of the reader. The writer is also less likely to make an error while explaining the plot to the reader.

Controversial Aspects of the Razor

Occam's razor is not an embargo against the positing of any kind of entity, or a recommendation of the simplest theory come what may.

The other things in question are the evidential support for the theory. Therefore, according to the principle, a simpler but less correct theory should not be preferred over a more complex but more correct one. It is this fact which gives the lie to the common misinterpretation of Occam's razor that "the simplest" one is usually the correct one.

For instance, classical physics is simpler than more recent theories; nonetheless it should not be preferred over them, because it is demonstrably wrong in certain respects.

Occam's razor is used to adjudicate between theories that have already passed "theoretical scrutiny" tests, and which are equally well-supported by the evidence. Furthermore, it may be used to prioritize empirical testing between two equally plausible but unequally testable hypotheses; thereby minimizing costs and wastes while increasing chances of falsification of the simpler-to-test hypothesis.

Another contentious aspect of the razor is that a theory can become more complex in terms of its structure (or syntax), while its ontology (or semantics) becomes simpler, or vice versa. Quine, in a discussion on definition, referred to these two perspectives as "economy of practical expression" and "economy in grammar and vocabulary", respectively. The theory of relativity is often given as an example of the proliferation of complex words to describe a simple concept.

Galileo Galilei lampooned the *misuse* of Occam's razor in his *Dialogue*. The principle is represented in the dialogue by *Simplicio*. The telling point that Galileo presented ironically was that if you really wanted to start from a small number of entities, you could always consider the letters of the alphabet as the fundamental entities, since you could certainly construct the whole of human knowledge out of them.

Anti-razors

Occam's razor has met some opposition from people who have considered it too extreme or rash. Walter of Chatton was a contemporary of William of Ockham (1287–1347) who took exception to Occam's razor and Ockham's use of it. In response he devised his own *anti-razor:* "If three things are not enough to verify an affirmative proposition about things, a fourth must be added, and so on." Although there have been a number of philosophers who have formulated similar anti-razors since Chatton's time, no one anti-razor has perpetuated in as much notoriety as Chatton's anti-razor, although this could be the case of the Late Renaissance Italian motto of unknown attribution *Se non θ vero, θ ben trovato* ("Even if it is not true, it is well conceived") when referred to a particularly artful explanation.

Anti-razors have also been created by Gottfried Wilhelm Leibniz (1646–1716), Immanuel Kant (1724–1804), and Karl Menger. Leibniz's version took the form of a principle of plenitude, as Arthur Lovejoyhas called it, the idea being that God created the most varied and populous of possible worlds. Kant felt a need to moderate the effects of Occam's razor and thus created his own counter-razor: "The variety of beings should not rashly be diminished."

Karl Menger found mathematicians to be too parsimonious with regard to variables so he formulated his Law Against Miserliness which took one of two forms: "Entities must not be reduced to the point of inadequacy" and "It is vain to do with fewer what requires more." See "Ockham's Razor and Chatton's Anti-Razor" (1984) by Armand Maurer. A less serious, but (some might say) even more extremist anti-razor is Pataphysics, the "science of imaginary solutions" invented by Alfred Jarry (1873–1907). Perhaps the ultimate in anti-reductionism, "Pataphysics seeks no less than to view each event in the universe as completely unique, subject to no laws but its own." Variations on this theme were subsequently explored by the Argentine writer Jorge Luis Borges in his story/mock-essay Tlφn, Uqbar, Orbis Tertius. There is also Crabtree's Bludgeon, which takes a cynical view that "[n]o set of mutually inconsistent observations can exist for which some human intellect cannot conceive a coherent explanation, however complicated."

Particle Filter

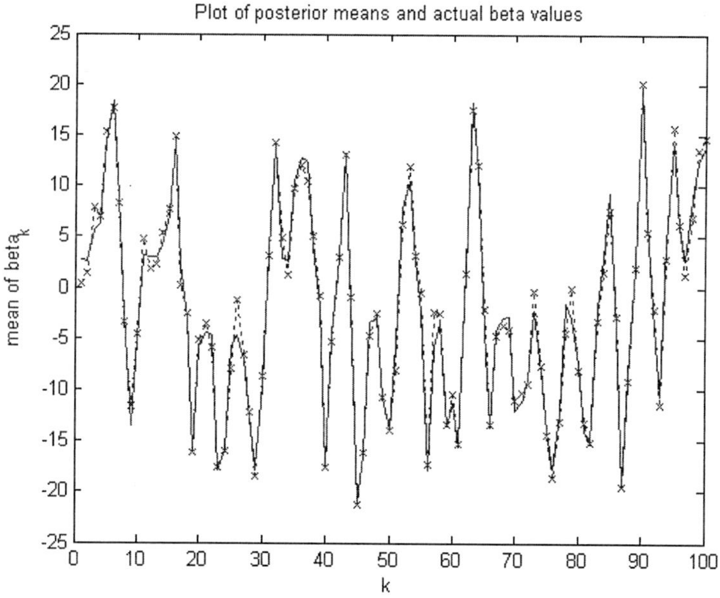

Figure: Result of particle filtering (red line) based on observed data generated from the blue line (Much larger image)

In statistics, particle filters, also known asSequential Monte Carlo methods (SMC), are sophisticated model estimation techniques based onsimulation. Particle filters have important applications in econometrics, and in other fields.

Particle filters are usually used to estimate Bayesian models in which the latent variables are connected in a Markov chain — similar to a hidden Markov model(HMM), but typically where the state space of the latent variables is continuous rather than discrete, and not sufficiently restricted to make exact inference tractable (as, for example, in a linear dynamical system, where the state space of the latent variables is restricted to Gaussian distributions and hence exact inference can be done efficiently using aKalman filter).

In the context of HMMs and related models, "filtering" refers to determining the distribution of a latent variable at a specific time, given all observations up to that time; particle filters are so named because they allow for approximate "filtering" (in the sense just given) using a set of "particles" (differently-weighted samples of the distribution).

Particle filters are the sequential ("on-line") analogue of Markov chain Monte Carlo (MCMC) batch methods and are often similar to importance sampling methods. Well-designed particle filters can often be much faster than MCMC. They are often an alternative to the Extended Kalman filter (EKF) orUnscented Kalman filter (UKF) with the advantage that, with sufficient samples, they approach the Bayesian optimal estimate, so they can be made more accurate than either the EKF or UKF. However, when the simulated sample is not sufficiently large, they might suffer from sample impoverishment. The approaches can also be combined by using a version of the Kalman filter as a proposal distribution for the particle filter.

Goal

The particle filter aims to estimate the sequence of hidden parameters, x_k for $k = 0,1,2,3,\ldots$, based only on the observed data y_k for $k = 0,1,2,3,\ldots$. All Bayesian estimates of x_k follow from the posterior distribution $p(x_k \mid y_0,y_1,\ldots,y_k)$. In contrast, the MCMC or importance sampling approach would model the full posterior $p(x_0,x_1,\ldots,x_k \mid y_0,y_1,\ldots,y_k)$.

Model

Particle methods assume x_k and the observations y_k can be modeled in this form:

- $x_0,x_1,\ldots$ is a first order Markov process such that $x_k \mid x_{k-1} \sim p_{x_k|x_{k-1}}(x \mid x_{k-1})$ and with an initial distribution $p(x_0)$.
- The observations $y_0,y_1,\ldots$ are conditionally independent provided that $x_0,x_1,\ldots$ are known.

In other words, each y_k only depends on x_k

$$y_k \mid x_k \sim p_{y|x}(y \mid x_k)$$

One example form of this scenario is

$$x_k = g(x_{k-1}) + w_k$$

$$y_k = h(x_k) + v_k$$

where both w_k and v_k are mutually independent and identically distributed sequences with knownprobability density functions and $g(\cdot)$ and $h(\cdot)$ are known functions. These two equations can be viewed as state space equations and look similar to the state space equations for the Kalman filter. If the functions $g(\cdot)$ and $h(\cdot)$ are linear, and if both w_k and v_k are Gaussian, the Kalman filter finds the exact Bayesian filtering distribution. If not, Kalman filter based methods are a first-order approximation (EKF) or a second-order approximation (UKF in general, but if probability distribution is Gaussian a third-order approximation is possible). Particle filters are also an approximation, but with enough particles they can be much more accurate.

Monte Carlo Approximation

Particle methods, like all sampling-based approaches (e.g., MCMC), generate a set of samples that approximate the filtering distribution $p(x_k \mid y_0,\ldots,y_k)$. So, with P samples, expectations with respect to the filtering distribution are approximated by

$$\int f(x_k)p(x_k \mid y_0,\ldots,y_k)dx_k \approx \frac{1}{P}\sum_{L=1}^{P} f(x_k^{(L)})$$

and $f(\cdot)$, in the usual way for Monte Carlo, can give all the moments etc. of the distribution up to some degree of approximation.

Sequential Importance Resampling (SIR)

Sequential importance resampling (SIR), the original particle filtering algorithm (Gordon et al. 1993), is a very commonly used particle filtering algorithm, which approximates the filtering distribution $p(x_k \mid y_0,\ldots,y_k)$ by a weighted set of P particles

$$\{(w_k^{(L)}, x_k^{(L)}) : L \in \{1,\ldots,P\}\}$$

The *importance weights* $w_k^{(L)}$ are approximations to the relative posterior probabilities (or densities) of the particles such that $\sum_{L=1}^{P} w_k^{(L)} = 1$

SIR is a sequential (i.e., recursive) version of importance sampling. As in importance sampling, the expectation of a function $f(\cdot)$ can be approximated as a weighted average

$$\int f(x_k)p(x_k \mid y_0,\ldots,y_k)dx_k \approx \sum_{L=1}^{P} w^{(L)} f(x_k^{(L)}).$$

For a finite set of particles, the algorithm performance is dependent on the choice of the *proposal distribution*

$$\pi(x_k \mid x_{0:k-1}, y_{0:k}) \,.$$

The *optimal proposal distribution* is given as the *target distribution*

$$\pi(x_k \mid x_{0:k-1}, y_{0:k}) = p(x_k \mid x_{k-1}, y_k).$$

However, the transition prior is often used as importance function, since it is easier to draw particles (or samples) and perform subsequent importance weight calculations:

$$\pi(x_k \mid x_{0:k-1}, y_{0:k}) = p(x_k \mid x_{k-1}).$$

Sequential Importance Resampling (SIR) filters with transition prior as importance function are commonly known as bootstrap filter and condensation algorithm.

Resampling is used to avoid the problem of degeneracy of the algorithm, that is, avoiding the situation that all but one of the importance weights are close to zero. The performance of the algorithm can be also affected by proper choice of resampling method. The *stratified sampling* proposed by Kitagawa (1996) is optimal in terms of variance.

A single step of sequential importance resampling is as follows:

1) For $L = 1,\ldots,P$ draw samples from the *proposal distribution*

$$x_k^{(L)} \sim \pi(x_k \mid x_{0:k-1}^{(L)}, y_{0:k})$$

2) For $L = 1,\ldots,P$ update the importance weights up to a normalizing constant:

$$\hat{w}_k^{(L)} = w_{k-1}^{(L)} \frac{p(y_k \mid x_k^{(L)})p(x_k^{(L)} \mid x_{k-1}^{(L)})}{\pi(x_k^{(L)} \mid x_{0:k-1}^{(L)}, y_{0:k})}.$$

Note that when we use the transition prior as the importance function, $\pi(x_k^{(L)} \mid x_{0:k-1}^{(L)}, y_{0:k}) = p(x_k^{(L)} \mid x_{k-1}^{(L)})$, this simplifies to the following :

$$\hat{w}_k^{(L)} = w_{k-1}^{(L)} p(y_k \mid x_k^{(L)}),$$

3) For $L=1,\dots,P$ compute the normalized importance weights:

$$w_k^{(L)} = \frac{\hat{w}_k^{(L)}}{\sum_{J=1}^{P} \hat{w}_k^{(J)}}$$

4) Compute an estimate of the effective number of particles as

$$\hat{N}_{eff} = \frac{1}{\sum_{L=1}^{P} \left(w_k^{(L)}\right)^2}$$

5) If the effective number of particles is less than a given threshold $\hat{N}_{eff} < N_{thr}$, then perform resampling:

a) Draw P particles from the current particle set with probabilities proportional to their weights. Replace the current particle set with this new one.

b) For $w_k^{(L)} = 1/P$. set $p_{x_k|y_{1:k}}(x \mid y_{1:k})$

The term *Sampling Importance Resampling* is also sometimes used when referring to SIR filters.

Sequential Importance Sampling (SIS)

- Is the same as Sequential Importance Resampling, but without the resampling stage.

"Direct version" Algorithm

The "direct version" algorithm is rather simple (compared to other particle filtering algorithms) and it uses composition and rejection. To generate a single sample x at k from $p_{x_k|y_{1:k}}(x \mid y_{1:k})$:

1) `Set n=0 (This will count the number of particles generated so far)
2) Uniformly choose an index L from the range $\{1,\dots,P\}$
3) Generate a test $\hat{x}$ from the distribution $p_{x_k|x_{k-1}}(x \mid x_{k-1|k-1}^{(L)})$
4) Generate the probability of $\hat{y}$ using $\hat{x}$ from $p_{y|x}(y_k \mid \hat{x})$ where y_k is the measured value
5) Generate another uniform u from $[0, m_k]$
6) Compare u and $p(\hat{y})$

6a) If u is larger then repeat from step 2

6b) If u is smaller then save $\hat{x}$ as $x_{k|k}^{(p)}$ and increment n.

7) If n == P then quit

The goal is to generate P "particles" at k using only the particles from k ” 1. This requires that a Markov equation can be written (and computed) to generate a x_k based only upon $x_{k\text{ ” }1}$. This algorithm uses composition of the P particles from k ” 1 to generate a particle at k and repeats (steps 2-6) until P particles are generated at k.

This can be more easily visualized if x is viewed as a two-dimensional array. One dimension is k and the other dimensions is the particle number. For example, $x(k,L)$ would be the Lth particle at k and can also be written $x_k^{(L)}$ (as done above in the algorithm). Step 3 generates a *potential* x_k based on a randomly chosen particle ($x_{k-1}^{(L)}$) at time k ” 1 and rejects or accepts it in step 6. In other words, thex_k values are generated using the previously generated $x_{k\text{ ” }1}$.

Artificial Immune System

In computer science, Artificial immune systems (AIS) are a class of computationally intelligent systems inspired by the principles and processes of the vertebrate immune system. The algorithms typically exploit the immune system's characteristics of learning and memory to solve a problem.

Definition

The field of Artificial Immune Systems (AIS) is concerned with abstracting the structure and function of the immune system to computational systems, and investigating the application of these systems towards solving computational problems from mathematics, engineering, and information technology. AIS is a sub-field of Biologically-inspired computing, and Natural computation, with interests inMachine Learning and belonging to the broader field of Artificial Intelligence.

Artificial Immune Systems (AIS) are adaptive systems, inspired by theoretical immunology and observed immune functions, principles and models, which are applied to problem solving.

AIS is distinct from computational immunology and theoretical biology that are concerned with simulating immunology using computational and mathematical models towards better understanding the immune system, although such models initiated the field of AIS and continue to provide a fertile ground for inspiration. Finally, the field of AIS is not concerned with the investigation of the immune system as a substrate computation, such as DNA computing.

History

AIS began in the mid 1980s with Farmer, Packard and Perelson's (1986) and Bersini and Varela's papers on immune networks (1990). However, it was only in the mid 90s that AIS became a subject area in its own right. Forrest *et al.* (on negative selection) and Kephart *et al.* published their first papers on AIS in 1994, and Dasgupta conducted extensive studies on Negative Selection Algorithms. Hunt and Cooke started the works on Immune Network models in 1995; Timmis and Neal continued this work and made some improvements. De Castro & Von Zuben's and Nicosia & Cutello's work (onclonal selection) became notable in 2002. The first book on Artificial Immune Systems was edited by Dasgupta in 1999.

New ideas, such as danger theory and algorithms inspired by the innate immune system, are also now being explored. Although some doubt that they are yet offering anything over and above existing AIS algorithms, this is hotly debated, and the debate is providing one the main driving forces for AIS development at the moment. Other recent developments involve the exploration of degeneracy in AIS models, which is motivated by its hypothesized role in open ended learning and evolution.

Originally AIS set out to find efficient abstractions of processes found in the immune system but, more recently, it is becoming interested in modelling the biological processes and in applying immune algorithms to bioinformatics problems.

In 2008, Dasgupta and Nino published a textbook on Immunological Computation which presents a compendium of up-to-date work related to immunity-based techniques and describes a wide variety of applications.

Techniques

The common techniques are inspired by specific immunological theories that explain the function and behavior of the mammalian adaptive immune system.

- Clonal Selection Algorithm: A class of algorithms inspired by the clonal selection theory of acquired immunity that explains how B and T lymphocytes improve their response to antigens over time called affinity maturation. These algorithms focus on the Darwinian attributes of the theory where selection is inspired by the affinity of antigen-antibody interactions, reproduction is inspired by cell division, and variation is inspired by somatic hypermutation. Clonal selection algorithms are most commonly applied to optimization and pattern recognition domains, some of which resemble parallel hill climbing and the genetic algorithm without the recombination operator.

- Negative Selection Algorithm: Inspired by the positive and negative selection processes that occur during the maturation of T cells in the thymus called T cell tolerance. Negative selection refers to the identification and deletion (apoptosis) of self-reacting cells, that is T cells that may select for and attack self tissues. This class of algorithms are typically used for classification and pattern recognition problem domains where the problem space is modeled in the complement of available knowledge. For example in the case of an anomaly detection domain the algorithm prepares a set of exemplar pattern detectors trained on normal (non-anomalous) patterns that model and detect unseen or anomalous patterns.
- Immune Network Algorithms: Algorithms inspired by the idiotypic network theory proposed by Niels Kaj Jerne that describes the regulation of the immune system by anti-idiotypic antibodies (antibodies that select for other antibodies). This class of algorithms focus on the network graph structures involved where antibodies (or antibody producing cells) represent the nodes and the training algorithm involves growing or pruning edges between the nodes based on affinity (similarity in the problems representation space). Immune network algorithms have been used in clustering, data visualization, control, and optimization domains, and share properties with artificial neural networks.
- Dendritic Cell Algorithms: The Dendritic Cell Algorithm (DCA) is an example of an immune inspired algorithm developed using a multi-scale approach. This algorithm is based on an abstract model ofdendritic cells (DCs). The DCA is abstracted and implemented through a process of examining and modeling various aspects of DC function, from the molecular networks present within the cell to the behaviour exhibited by a

population of cells as a whole. Within the DCA information is granulated at different layers, achieved through multi-scale processing.

Evolutionary Computation

In computer science, evolutionary computation is a subfield of artificial intelligence (more particularly computational intelligence) that involves combinatorial optimization problems.

Evolutionary computation uses iterative progress, such as growth or development in a population. This population is then selected in a guided random search using parallel processing to achieve the desired end. Such processes are often inspired by biological mechanisms of evolution.

As evolution can produce highly optimised processes and networks, it has many applications incomputer science. Here, simulations of evolution using evolutionary algorithms and artificial life started with the work of Nils Aall Barricelli in the 1960s, and was extended by Alex Fraser, who published a series of papers on simulation of artificial selection. Artificial evolution became a widely recognised optimisation method as a result of the work of Ingo Rechenberg in the 1960s and early 1970s, who used evolution strategies to solve complex engineering problems. Genetic algorithms in particular became popular through the writing of John Holland. As academic interest grew, dramatic increases in the power of computers allowed practical applications, including the automatic evolution of computer programs. Evolutionary algorithms are now used to solve multi-dimensional problems more efficiently than software produced by human designers, and also to optimise the design of systems.

History

The use of Darwinian principles for automated problem solving originated in the fifties. It was not until the sixties that three distinct interpretations of this idea started to be developed in three different places.

Evolutionary programming was introduced by Lawrence J. Fogel in the USA, while John Henry Hollandcalled his method a genetic algorithm. In Germany Ingo Rechenberg and Hans-Paul Schwefelintroduced evolution strategies. These areas developed separately for about 15 years. From the earlynineties on they are unified as different representatives ("dialects") of one technology, called evolutionary computing. Also in the early nineties, a fourth stream following the general ideas had emerged – genetic programming. Since the 1990s, evolutionary computation has largely become swarm-based computation, and nature-inspired algorithms are becoming an increasingly significant part.

These terminologies denote the field of evolutionary computing and consider evolutionary programming, evolution strategies, genetic algorithms, and genetic programming as sub-areas.

Techniques

Evolutionary computing techniques mostly involve metaheuristic optimization algorithms. Broadly speaking, the field includes:

Evolutionary algorithms

- Genetic algorithms
- Evolutionary programming
- Evolution strategy
- Eagle strategy
- Swarm intelligence
- Ant colony optimization
- Particle swarm optimization

and in a lesser extent also:

- Artificial life (also see digital organism)
- Artificial immune systems
- Cuckoo search
- Cultural algorithms
- Differential evolution
- Firefly algorithm
- Harmony search
- Learning classifier systems
- Learnable Evolution Model
- Parallel simulated annealing
- Self-organization such as self-organizing maps, competitive learning
- Self-Organizing Migrating Genetic Algorithm
- Swarm-based computing

Evolutionary Algorithms

Evolutionary algorithms form a subset of evolutionary computation in that they generally only involve techniques implementing mechanisms inspired by biological evolution such as reproduction, mutation,recombination, natural selection and survival of the fittest. Candidate solutions to the optimization problem play the role of individuals in a population, and the cost function determines the environment within which the solutions "live" (see also fitness function). Evolution of the population then takes place after the repeated application of the above operators.

In this process, there are two main forces that form the basis of evolutionary systems:Recombination and mutation create the necessary diversity and thereby facilitate novelty, whileselection acts as a force increasing quality.

Many aspects of such an evolutionary process are stochastic. Changed pieces of information due to recombination and mutation are randomly chosen. On the other hand, selection operators can be either deterministic, or stochastic. In the latter case, individuals with a higher fitness have a higher chance to be selected than individuals with a lower fitness, but typically even the weak individuals have a chance to become a parent or to survive.

Evolutionary Computation Practitioners

- Kalyanmoy Deb
- David E. Goldberg
- John Henry Holland
- John Koza
- Peter Nordin
- Ingo Rechenberg
- Hans-Paul Schwefel
- Peter J. Fleming
- Carlos M. Fonseca
- Lee Graham.

Evolutionary Programming

Evolutionary programming is one of the four major evolutionary algorithm paradigms. It is similar togenetic programming, but the structure of the program to be optimized is fixed, while its numerical parameters are allowed to evolve. It was first used by Lawrence J. Fogel in the US in 1960 in order to use simulated evolution as a learning process aiming to generate artificial intelligence. Fogel used finite state machines as predictors and evolved them. Currently evolutionary programming is a wide evolutionary computingdialect with no fixed structure or (representation), in contrast with some of the other dialects. It is becoming harder to distinguish from evolutionary strategies.

Its main variation operator is mutation; members of the population are viewed as part of a specific species rather than members of the same species therefore each parent generates an offspring, using a $(\mu + \mu)$ survivor selection.

Evolution Strategy

In computer science, evolution strategy (ES) is an optimization technique based on ideas of adaptation and evolution. It belongs to the general class of evolutionary computation or artificial evolution methodologies.

History

The evolution strategy optimization technique was created in the early 1960s and developed further in the 1970s and later by Ingo Rechenberg, Hans-Paul Schwefel and his co-workers.

Methods

Evolution strategies use natural problem-dependent representations, and primarily mutation andselection, as search operators. In common with evolutionary algorithms, the operators are applied in a loop. An iteration of the loop is called a generation. The sequence of generations is continued until a termination criterion is met.

As far as real-valued search spaces are concerned, mutation is normally performed by adding anormally distributed random value to each vector component. The step size or mutation strength (i.e. the standard deviation of the normal distribution) is often governed by self-adaptation (see evolution window). Individual step sizes for each coordinate or correlations between coordinates are either governed by self-adaptation or by covariance matrix adaptation (CMA-ES).

The (environmental) selection in evolution strategies is deterministic and only based on the fitness rankings, not on the actual fitness values. The resulting algorithm is therefore invariant with respect to monotonic transformations of the objective function. The simplest evolution strategy operates on a population of size two: the current point (parent) and the result of its mutation.

Only if the mutant's fitness is at least as good as the parent one, it becomes the parent of the next generation. Otherwise the mutant is disregarded.

This is a *(1 + 1)-ES*. More generally, λ mutants can be generated and compete with the parent, called *(1 + λ)-ES*. In (1 , λ)-ES the best mutant becomes the parent of the next generation while the current parent is always disregarded.

For some of these variants, proofs oflinear convergence (in a stochastic sense) have been derived on unimodal objective functions.

Contemporary derivatives of evolution strategy often use a population of μ parents and also recombination as an additional operator, called *(μ/ρ+, λ)-ES*. This makes them less prone to get stuck in local optima.

Eagle Strategy

Eagle strategy is a search strategy for solving nonlinear optimization problems, and this strategy was developed by Xin-she Yang and Suash Deb, based on the foraging behaviour of eagle species such asGolden Eagles.

It is a two-stage iterative strategy. It uses a slow search stage and a fast stage to simulate an eagle searching for prey tends to search on a large area and then quickly switches to a rapid chasing phase once a prey is in sight. In optimization, it uses a coarse search stage on a larger area in a search space in combination with an intensive faster search algorithm in the neighbourhood of promising solutions. Two stages interchanges and proceed iteratively.

As there are two stage, each stage can employ different algorithms. Studies show that such a combination is better than any of its components.

In the simplest case, when the fist stage does not use any algorithm (just initialization), it essentially degenerates into a hill-climbing with random restart. However, this strategy could be potentially much more powerful if a good combination of different algorithms is used.

Collective Intelligence

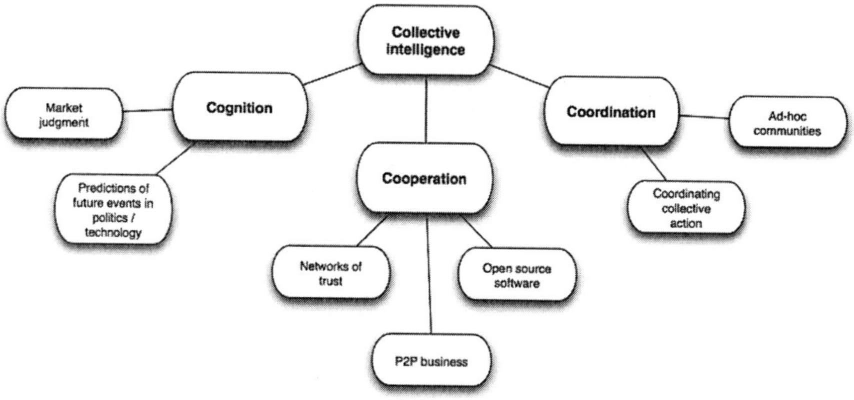

Figure: Types of collective intelligence

Collective intelligence is a shared or group intelligence that emerges from the collaboration and competition of many individuals and appears in consensus decision making in bacteria, animals, humans and computer networks.

The idea emerged from the writings of Douglas Hofstadter (1979), Peter Russell (1983), Tom Atlee (1993),Pierre Lévy (1994), Howard Bloom(1995), Francis Heylighen (1995), Douglas Engelbart, Cliff Joslyn, Ron Dembo, Gottfried Mayer-Kress(2003) and other theorists. Collective intelligence is referred to as Symbiotic intelligence by Norman Lee Johnson. The concept is relevant in sociology, business, computer science and mass communications: it also appears in science fiction, frequently in the form of telepathically-linked species and cyborgs.

History

A precursor of the concept is found in entomologist William Morton Wheeler's observation that seemingly independent individuals can cooperate so closely as to become indistinguishable from a single organism (1911). Wheeler saw this collaborative process at work in ants that acted like the cells of a single beast he called a "superorganism".

In 1912 Émile Durkheim identified society as the sole source of human logical thought. He argued, in "The Elementary Forms of Religious Life" that society constitutes a higher intelligence because it transcends the individual over space and time. Other antecedents are Vladimir Vernadsky's concept of "noosphere" and H.G. Wells's concept of "world brain" (see also the term "global brain"). Peter Russell, Elisabet Sahtouris, and Barbara Marx Hubbard (originator of the term "conscious evolution") are inspired by the visions of a noosphere — a transcendent, rapidly evolving collective intelligence — an informational cortex of the planet. The notion has more recently been examined by the philosopher Pierre Lévy.

Dimensions

Howard Bloom has discussed mass behavior - collective behavior from the level of quarks to the level of bacterial, plant, animal, and human societies. He stresses the biological adaptations that have turned most of this earth's living beings into components of what he calls "a learning machine". In 1986 Bloom combined the concepts of apoptosis, parallel distributed processing, group selection, and the superorganism to produce a theory of how collective intelligence works. Later he showed how the collective intelligences of competing bacterial colonies and human societies can be explained in terms of computer-generated "complex adaptive systems" and the "genetic algorithms", concepts pioneered by John Holland.

Bloom traced the evolution of collective intelligence to our bacterial ancestors 1 billion years ago and demonstrated how a multi-species intelligence has worked since the beginning of life. Ant societies exhibit more intelligence, in terms of technology, than any other animal except for humans and co-operate in keeping livestock, for example aphids for "milking". Leaf cutters care for fungi and carry leaves to feed the fungi.

David Skrbina cites the concept of a 'group mind' as being derived from Plato's concept ofpanpsychism (that mind or consciousness is omnipresent and exists in all matter). He develops the concept of a 'group mind' as articulated by Thomas Hobbes in "Leviathan" and Fechner's arguments for a collective consciousness of mankind. He cites Durkheim as the most notable advocate of a 'collective consciousness" and Teilhard de Chardin as a thinker who has developed the philosophical implications of the group mind.

Tom Atlee focuses primarily on humans and on work to upgrade what Howard Bloom calls "the group IQ". Atlee feels that collective intelligence can be encouraged "to overcome 'groupthink' and individualcognitive bias in order to allow a collective to cooperate on one process—while achieving enhanced intellectual performance."

George Pór defined the collective intelligence phenomenon as "the capacity of human communities to evolve towards higher order complexity and harmony, through such innovation mechanisms as differentiation and integration, competition and collaboration." Atlee and Pór state that "collective intelligence also involves achieving a single focus of attention and standard of metrics which provide an appropriate threshold of action". Their approach is rooted in Scientific Community Metaphor.

Atlee and Pór suggest that the field of collective intelligence should primarily be seen as a humanenterprise in which mind-sets, a willingness to share and an openness to the value of distributed intelligence for the common good are paramount, though group theory and artificial intelligence have something to offer. Individuals who respect collective intelligence are confident of their own abilities and recognize that the whole is indeed greater than the sum of any individual parts.Maximizing collective intelligence relies on the ability of an organization to accept and develop "The Golden Suggestion", which is any potentially useful input from any member. Groupthink often hampers collective intelligence by limiting input to a select few individuals or filtering potential Golden Suggestions without fully developing them to implementation.

Robert David Steele Vivas in *The New Craft of Intelligence* portrayed all citizens as "intelligence minutemen," drawing only on legal and ethical sources of information, able to create a "public intelligence" that keeps public officials and corporate managers honest, turning the concept of "national intelligence" (previously concerned about spies and secrecy) on its head.

According to Don Tapscott and Anthony D. Williams, collective intelligence is mass collaboration. In order for this concept to happen, four principles need to exist;

Openness: Sharing ideas and intellectual property: though these resources provide the edge over competitors more benefits accrue from allowing others to share ideas and gain significant improvement and scrutiny through collaboration.

Peering: Horizontal organization as with the 'opening up' of the Linux program where users are free to modify and develop it provided that they make it available for others. Peering succeeds because it encourages self-organization – a style of production that works more effectively than hierarchical management for certain tasks.

Sharing: Companies have started to share some ideas while maintaining some degree of control over others, like potential and critical patent rights. Limiting all intellectual property shuts out opportunities, while sharing some expands markets and brings out products faster.

Acting Globally: The advancement in communication technology has prompted the rise of global companies at low overhead costs. The internet is widespread, therefore a globally integrated company has no geographical boundaries and may access new markets, ideas and technology.

Examples

Political parties mobilize large numbers of people to form policy, select candidates and finance and run election campaigns. Knowledge focusing through various voting methods allows perspectives to converge through the assumption that uninformed voting is to some degree random and can be filtered from the decision process leaving only a residue of informed consensus. Critics point out that often bad ideas, misunderstandings, and misconceptions are widely held, and that structuring of the decision process must favor experts who are presumably less prone to random or misinformed voting in a given context.

Military units, trade unions, and corporations satisfy some definitions of CI — the most rigorous definition would require a capacity to respond to very arbitrary conditions without orders or guidance from "law" or "customers" to constrain actions. Online advertising companies are using collective intelligence to bypass traditional marketing and creative agencies.

In Learner generated context a group of users marshal resources to create an ecology that meets their needs often (but not only) in relation to the co-configuration, co-creation and co-design of a particular learning space that allows learners to create their own context. Learner generated contexts represent an *ad hoc* community that facilitates coordination of collective action in a network of trust. An example of Learner generated context is found on the Internet when collaborative users pool knowledge in a "shared intelligence space" such as Wikipedia. As the Internet has developed so has the concept of CI as a shared public forum. The global accessibility and availability of the Internet has allowed more people than ever to contribute and access ideas. (Flew 2008)

Improvisational actors also experience a type of collective intelligence which they term 'Group Mind'. A further example of collective intelligence is found in idea competitions.

Mathematical Techniques

One measure sometimes applied, especially by more artificial intelligence focused theorists, is a "collective intelligence quotient" (or "cooperation quotient")—which presumably can be measured like the "individual" intelligence quotient (IQ)—thus making it possible to determine the marginal extra intelligence added by each new individual participating in the collective, thus using metrics to avoid the hazards of group think and stupidity.

In 2001, Tadeusz (Ted) Szuba from the AGH University in Poland proposed a formal model for the phenomenon of collective intelligence. It is assumed to be an unconscious, random, parallel, and distributed computational process, run in mathematical logic by the social structure.

In this model, beings and information are modeled as abstract information molecules carrying expressions of mathematical logic. They are quasi-randomly displacing due to their interaction with their environments with their intended displacements.

Their interaction in abstract computational space creates multi-thread inference process which we perceive as collective intelligence. Thus, a non-Turingmodel of computation is used.

This theory allows simple formal definition of collective intelligence as the property of social structure and seems to be working well for a wide spectrum of beings, from bacterial colonies up to human social structures. Collective intelligence considered as a specific computational process is providing a straightforward explanation of several social phenomena.

For this model of collective intelligence, the formal definition of IQS (IQ Social) was proposed and was defined as "the probability function over the time and domain of N-element inferences which are reflecting inference activity of the social structure." While IQS seems to be computationally hard, modeling of social structure in terms of a computational process as described above gives a chance for approximation. Prospective applications are optimization of companies through the maximization of their IQS, and the analysis of drug resistance against collective intelligence of bacterial colonies.

Digital Media

New media are often associated with the promotion and enhancement of collective intelligence. The ability of new media to easily store and retrieve information, predominantly through databases and the Internet, allows for it to be shared without difficulty. Thus, through interaction with new media, knowledge easily passes between sources (Flew 2008) resulting in a form of collective intelligence. The use of interactive new media, particularly the internet, promotes online interaction and this distribution of knowledge between users.

Francis Heylighen, Valerie Turchin, and Gottfried Mayer-Kress are among those who view collective intelligence through the lens of computer science and cybernetics. Collective intelligence can be defined as a form of networking enabled by the internet. The developer of the World Wide Web, Tim Berners-Lee, aimed to promote sharing and publishing of information globally. Later his employer opened up the technology for free use.

In the early '90s, the Internet's potential was still untapped, until the mid 1990s when 'critical mass', as termed by the head of the Advanced Research Project Agency (ARPA), Dr. J.C.R. Licklider, demanded more accessibility and utility. The driving force of this form of collective intelligence is the digitization of information and communication. Henry Jenkins, a key theorist of new media and media convergence draws on the theory that collective intelligence can be attributed to media convergence and participatory culture (Flew 2008). Collective intelligence is not merely a quantitative contribution of information from all cultures, it is also qualitative.

Levy and de Kerckhove consider CI from a mass communications perspective, focusing on the ability of networked ICT's to enhance the community knowledge pool. They suggest that these communications tools enable humans to interact and to share and collaborate with both ease and speed (Flew 2008). With the development of the Internet and its widespread use, the opportunity to contribute to community-based knowledge forums, such as Wikipedia, is greater than ever before. These computer networks give participating users the opportunity to store and to retrieve knowledge through the collective access to these databases and allow them to "harness the hive" (Raymond 1998; Herz 2005 in Flew 2008). Researchers at the MIT Center for Collective Intelligence research and explore collective intelligence of groups of people and computers.

In this context collective intelligence is often confused with shared knowledge. The former is knowledge that is generally available to all members of a community while the latter is information known by all members of a community. Collective intelligence as represented by Web 2.0 has less user engagement than collaborative intelligence. An art project using Web 2.0 platforms is "Shared Galaxy", an experiment developed by an anonymous artist to create a collective identity that shows up as one person on several platforms like MySpace, Facebook, Youtube and Second Life. The password is written in the profiles and the accounts named "Shared Galaxy" are open to be used by anyone. In this way many take part in being one.

It has been argued that media, particularly central media, cannot promote intelligence, due to the inherent inability of Central media to adequately deal with complex issues such as the Environmental Crisis. See The IRG Solution - hierarchical incompetence and how to overcome it1984, argued, that Central media and government type hierarchical organizations. The book argued that collective intelligence could only emerge from vast informal networks of human interaction, something which Media do not promote.

Growth of the Internet and mobile telecom has also produced "swarming" or "rendezvous" events that enable meetings or even dates on demand. The full impact has yet to be felt but the anti-globalization movement, for example, relies heavily on e-mail, cell phones, pagers, SMS and other means of organizing. Atlee discusses the connections between these events and the political views that drive them. The Indymedia organization does this in a more journalistic way. Such resources could combine into a form of collective intelligence accountable only to the current participants yet with some strong moral or linguistic guidance from generations of contributors - or even take on a more obviously democratic form to advance shared goals.

Social Bookmarking

In social bookmarking (also called collaborative tagging), users assign tags to resources shared with other users, which gives rise to a type of information organisation that emerges from thiscrowdsourcing process. The resulting information structure can be seen as reflecting the collective knowledge (or collective intelligence) of a community of users and is commonly called a "Folksonomy".

Recent research using data from the social bookmarking website Del.icio.us, has shown that collaborative tagging systems exhibit a form of complex systems (or self-organizing) dynamics.Although there is no central controlled vocabulary to constrain the actions of individual users, the distributions of tags that describe different resources has been shown to converge over time to a stablepower law distributions. Once such stable distributions form, examining the correlations between different tags can be used to construct simple folksonomy graphs, which can be efficiently partitioned to obtained a form of community or shared vocabularies. Such vocabularies can be seen as a form of collective intelligence, emerging from the decentralised actions of a community of users.

Video Games

Games such as *The Sims* Series, and *Second Life* are designed to be non-linear and to depend on collective intelligence for expansion. This way of sharing is gradually evolving and influencing the mindset of the current and future generations. For them, collective intelligence has become a norm. In Terry Flew's discussion of 'interactivity' in the online games environment, the ongoing interactive dialogue between users and game developers, he refers to Pierre Levy's concept of Collective Intelligence (Levy 1998) and argues this is active in videogames as clans or guilds in MMORPGconstantly work to achieve goals. Henry Jenkins proposes that the participatory cultures emerging between games producers, media companies, and the end-users mark a fundamental shift in the nature of media production and consumption. Jenkins argues that this new participatory culture arises at the intersection of three broad new media trends. Firstly, the development of new media tools/technologies enabling the creation of content. Secondly, the rise of subcultures promoting such creations, and lastly, the growth of value adding media conglomerates, which foster image, idea and narrative flow. Cultural theorist and online community developer, John Banks considered the contribution of online fan communities in the creation of the Trainz product. He argued that its commercial success was fundamentally dependant upon "the formation and growth of an active and vibrant online fan community that would both actively promote the product and create content-extensions and additions to the game software".

The increase in user created content and interactivity gives rise to issues of control over the game itself and ownership of the player-created content. This gives rise to fundamental legal issues, highlighted by Lessig and Bray and Konsynski, such as Intellectual Property and property ownership rights. Gosney extends this issue of Collective Intelligence in videogames one step further in his discussion of Alternate Reality Gaming. This genre, he describes as an "across-media game that deliberately blurs the line between the in-game and out-of-game experiences" as events that happen outside the game reality "reach out" into the player's lives in order to bring them together. Solving the game requires "the collective and collaborative efforts of multiple players"; thus the issue of collective and collaborative team play is essential to ARG. Gosney argues that the Alternate Reality genre of gaming dictates an unprecedented level of collaboration and "collective intelligence" in order to solve the mystery of the game.

Stock Market Predictions

Because of the Internet's ability to rapidly convey large amounts of information throughout the world, the use of collective intelligence to predict stock prices and stock price direction has become increasingly viable. Websites aggregate stock market information that is as current as possible so professional or amateur stock analysts can publish their viewpoints, enabling amateur investors to submit their financial opinions and create an aggregate opinion. The opinion of all investor can be weighed equally so that a pivotal premise of the effective application of collective intelligence can beapplied: the masses, including a broad spectrum of stock market expertise, can be utilized to more accurately predict the behavior of financial markets.

Collective intelligence underpins the efficient-market hypothesis of Eugene Fama - although the term collective intelligence is not used explicitly in his paper. Fama cites research conducted byMichael Jensen in which 89 out of 115 selected funds underperformed relative to the index during the period from 1955 to 1964. But after removing the loading charge (up-front fee) only 72 underperformed while after removing brokerage costs only 58 underperformed. On the basis of such evidence index funds became popular investment vehicles using the collective intelligence of the market, rather than the judgement of professional fund managers, as an investment strategy.

Views

Tom Atlee reflects that, although humans have an innate ability to gather and analyze data, they are affected by culture, education and social institutions. A single person tends to make decisions motivated by self-preservation. In addition, humans lack a way to make choices that balance innovation and reality. Therefore, without collective intelligence, humans may drive themselves into extinction based on their selfish needs.

Phillip Brown and Hugh Lauder quotes Bowles and Gintis (1976) that in order to truly define collective intelligence, it is crucial to separate 'intelligence' from IQism. They go on to argue that intelligence is an achievement and can only be developed if allowed to. For example, earlier on, groups from the lower levels of society are severely restricted from aggregating and pooling their intelligence. This is because the elites fear that the collective intelligence would convince the people to rebel. If there is no such capacity and relations, there would be no infrastructure on which collective intelligence is built (Brown & Lauder 2000, p. 230). This reflects how powerful collective intelligence can be if left to develop.

Research performed by Tapscott and Williams has provided a few examples of the benefits of collective intelligence to business:

Talent Utilization: At the rate technology is changing, no firm can fully keep up in the innovations needed to compete. Instead, smart firms are drawing on the power of mass collaboration to involve participation of the people they could not employ.

Demand Creation: Firms can create a new market for complementary goods by engaging in open source community.

Costs Reduction: Mass collaboration can help to reduce costs dramatically. Firms can release a specific software or product to be evaluated or debugged by online communities. The results will be more personal, robust and error-free products created in a short amount of time and costs.

Skeptics, especially those critical of artificial intelligence and more inclined to believe that risk ofbodily harm and bodily action are the basis of all unity between people, are more likely to emphasize the capacity of a group to take action and withstand harm as one fluid mass mobilization, shrugging off harms the way a body shrugs off the loss of a few cells. This strain of thought is most obvious in the anti-globalization movement and characterized by the works of John Zerzan, Carol Moore, andStarhawk, who typically shun academics. These theorists are more likely to refer to ecological and collective wisdom and to the role of consensus process in making ontological distinctions than to any form of "intelligence" as such, which they often argue does not exist, or is mere "cleverness".

Harsh critics of artificial intelligence on ethical grounds are likely to promote collective wisdom-building methods, such as the new tribalists and the Gaians. Whether these can be said to be collective intelligence systems is an open question. Some, e.g. Bill Joy, simply wish to avoid any form of autonomous artificial intelligence and seem willing to work on rigorous collective intelligence in order to remove any possible niche for AI.

Cuckoo Search

Cuckoo search (CS) is an optimization algorithm developed by Xin-she Yang and Suash Deb in 2009. It was inspired by the obligate brood parasitism of some cuckoo species by laying their eggs in the nests of other host birds (of other species). Some host birds can engage direct conflict with the intruding cuckoos.

For example, if a host bird discovers the eggs are not their own, it will either throw these alien eggs away or simply abandon its nest and build a new nest elsewhere.

Some cuckoo species such as the New World brood-parasitic Tapera have evolved in such a way that female parasitic cuckoos are often very specialized in the mimicry in colors and pattern of the eggs of a few chosen host species

Cuckoo search idealized such breeding behavior, and thus can be applied for various optimization problems. It seems that it can outperform other metaheuristic algorithms in applications.

Another seemingly unrelated algorithm for a completely different area of applications is called cuckoo hashing which was developed by Rasmus Pagh and Fleming Friche Rodler in 2001.

Cuckoo Search

Cuckoo Search (CS) Uses the Following Representations: Each egg in a nest represents a solution, and a cuckoo egg represents a new solution. The aim is to use the new and potentially better solutions (cuckoos) to replace a not-so-good solution in the nests. In the simplest form, each nest has one egg. The algorithm can be extended to more complicated cases in which each nest has multiple eggs representing a set of solutions.

CS is based on three idealized rules:

1. Each cuckoo lays one egg at a time, and dumps its egg in a randomly chosen nest;

2. The best nests with high quality of eggs will carry over to the next generation;
3. The number of available hosts nests is fixed, and the egg laid by a cuckoo is discovered by the host bird with a probability $p_a \in (0,1)$. Discovering operate on some set of worst nests, and discovered solutions dumped from farther calculations.

In addition, Yang and Deb discovered that the random-walk style search is better performed by Lévy flights rather than simple random walk.

The pseudo-code can be summarized as:

Objective function: $f(\mathbf{x})$, $\mathbf{x} = (x_1, x_2, \ldots, x_d)$;

Generate an initial population of n host nests;

While (t<MaxGeneration) or (stop criterion)

Get a cuckoo randomly (say, i) and replace its solution by performing Lévy flights;

Evaluate its quality/fitness F_i

[For maximization, $F_i \propto f(\mathbf{x}_i)$];

Choose a nest among n (say, j) randomly;

if ($F_i > F_j$),

Replace j by the new solution;

end if

A fraction (p_a) of the worse nests are abandoned and new ones are built;

Keep the best solutions/nests;

Rank the solutions/nests and find the current best;

Pass the current best solutions to the next generation;

end while

Post-processing the results and visualization;

An important advantage of this algorithm is its simplicity. In fact, comparing with other population- or agent - ased metaheuristic algorithms such as particle swarm optimization and harmony search, there is essentially only a single parameter p_a in CS (apart from the population size n).

Therefore, it is very easy to implement.

Random Walks and the Step Size

An important issue is the applications of Levy flights and random walks in the generic equation for generating new solutions

$$\mathbf{x}_{t+1} = \mathbf{x}_t + sE_t,$$

where E_t is drawn from a standard normal distribution with zero mean and unity standard deviation for random walks, or drawn from Levy distribution for Levy flights. Obviously, the random walks can also be linked with the similarity between a cuckoo's egg and the host's egg which can be tricky in implementation. Here the step size s determines how far a random walker can go for a fixed number of iterations. The generation of Levy step size is often tricky, and a good algorithm is Mantegna's algorithm.

If s is too large, then the new solution generated will be too far away from the old solution (or even jump out side of the bounds). Then, such a move is unlikely to be accepted. If s is too small, the change is too small to be significant, and consequently such search is not efficient. So a proper step size is important to maintain the search as efficient as possible.

As an example, for simple isotropic random walks, we know that the average distance r traveled in the d-dimension space is

$$r^2 = 2dDt,$$

where $D = s^2 / 2\tau$ is the effective diffusion coefficient. Here s is the step size or distance traveled at each jump, and τ is the time taken for each jump. The above equation implies that

$$s^2 = \frac{\tau r^2}{t d}.$$

For a typical length scale L of a dimension of interest, the local search is typically limited in a region of $r = L / 10$. For $\tau = 1$ and t=100 to 1000, we have $s \approx 0.01L$ for d=1, and $s \approx 0.001L$ for d=10. Therefore, we can use s/L=0.001 to 0.01 for most problems. Though the exact derivation may require detailed analysis of the behaviour of Levy flights. Algorithm and convergence analysis will be fruitful, because there are many open problems related to metaheuristics

Implementation

The pseudo code was given in a sequential form, but Yang and Deb provides a vectorized implementation which is very efficient. In the real world, if a cuckoo's egg is very similar to a host's eggs, then this cuckoo's egg is less likely to be discovered, thus the fitness should be related to the difference in solutions. Therefore, it is a good idea to do a random walk in a biased way with some random step sizes. For Matlab implementation, this biased random walk can partly be achieved by

stepsize=rand*(nest(randperm(n),:)-nest(randperm(n),:));

new_nest=nest+stepsize.*K;

where K=rand(size(nest))>pa and pa is the discovery rate. A standard CS matlab can be found here

An object-oriented software implementation of cuckoo search was provided by Bacanin On the other hand, a modified cuckoo search algorithm is also implemented for unconstrained optimization problems.

An open source implementation of Modified Cuckoo Search can be found herehttp://code.google.com/p/modified-cs/

Modified Cuckoo Search

A modification of the standard Cuckoo Search was made by Walton et al. with the aim to speed up convergence. The modification involves the additional step of information exchange between the top eggs. It was shown that Modified Cuckoo Search (MCS) outperforms the standard cuckoo search and other algorithms, in terms of convergence rates, when applied to a series of standard optimization benchmark objective functions.

Subsequently, the modified cuckoo search has been applied to optimize unstructured mesh which reduces computational cost significantly. In addition, another interesting improvement to cuckoo search is the so-called quantum-inspired cuckoo search with convincing results

Multiobjective Cuckoo Search (MOCS)

A multiobjective cuckoo search (MOCS) method has been formulated to deal with multi-criteria optimization problems. This approach uses random weights to combine multiple objectives to a single objective. As the weights vary randomly, Pareto fronts can be found so the points can distributed diversely over the fronts.

Hybridization

Though cuckoo search is a swarm-intelligence-based algorithm, it can be still hybridized with other swarm-based algorithms such as

PSO. For example, a hybrid CS-PSO algorithm seems to remedy the defect of PSO.

Applications

The applications of Cuckoo Search into engineering optimization problems have shown its promising efficiency. For example, for both spring design and welded beam design problems, CS obtained better solutions than existing solutions in literature. A promising discrete cuckoo search algorithm is recently proposed to solve nurse scheduling problem. An efficient computation approach based on cuckoo search has been proposed for data fusion in wireless sensor networks. A new quantum-inspired cuckoo search was developed to solve Knapsack problems, which shows its effectiveness.

A conceptual comparison of the cuckoo search with PSO, DE and ABC suggest that CS and DE algorithms provide more robust results than PSO and ABC. An extensive detailed study of various structural optimization problems suggests that cuckoo search obtains better results than other algorithms. In addition, a new software testing approach has been developed based on cuckoo search. In addition, cuckoo search is particularly suitable for large scale problems, as shown in a recent study. Cuckoo search has been applied to train neural networks with improved performance. Furthermore, CS is successfully applied to train spiking neural models. Cuckoo search has also been used to optimize web service composition process and planning graphs.

Cuckoo search is a reliable approach for embedded system design and design optimization. An interesting application of cuckoo search is to solve boundary value problems.

Chapter 5

Computational Complexity Theory

Computational complexity theory is a branch of the theory of computation in theoretical computer science and mathematics that focuses on classifying computational problems according to their inherent difficulty, and relating those classes to each other. In this context, a computational problem is understood to be a task that is in principle amenable to being solved by a computer (which basically means that the problem can be stated by a set of mathematical instructions). Informally, a computational problem consists of problem instances and solutions to these problem instances. For example, primality testing is the problem of determining whether a given number is prime or not. The instances of this problem are natural numbers, and the solution to an instance is *yes* or *no* based on whether the number is prime or not.

A problem is regarded as inherently difficult if its solution requires significant resources, whatever the algorithm used. The theory formalizes this intuition, by introducing mathematical models of computation to study these problems and quantifying the amount of resources needed to solve them, such as time and storage. Other complexity measures are also used, such as the amount of communication (used in communication complexity), the number of gates in a circuit (used in circuit complexity) and the number of processors (used in parallel computing). One of the roles of computational complexity theory is to determine the practical limits on what computers can and cannot do.

Closely related fields in theoretical computer science are analysis of algorithms and computability theory. A key distinction between analysis of algorithms and computational complexity theory is that the former is devoted to analyzing the amount of resources needed by a particular algorithm to solve a problem, whereas the latter asks a more general question about all possible algorithms that could be used to solve the same problem. More precisely, it tries to classify problems that can or cannot be solved with appropriately restricted resources. In turn, imposing restrictions on the available resources is what distinguishes computational complexity from computability theory: the latter theory asks what kind of problems can, in principle, be solved algorithmically.

Computational Problem

In theoretical computer science, a computational problem is a mathematical object representing a collection of questions that computers might want to solve. For example, the problem of factoring "Given a positive integer n, find a nontrivial prime factor of n.– is a computational problem. Computational problems are one of the main objects of study in theoretical computer science. The field of algorithms studies methods of solving computational problems efficiently. The complementary field of computational complexity attempts to explain why certain computational problems are intractable for computers.

A computational problem can be viewed as an infinite collection of *instances* together with a *solution*for every instance. For example in the factoring problem, the instances are the integers n, and solutions are prime numbers p that describe nontrivial prime factors of n.

It is conventional to represent both instances and solutions by binary strings, namely elements of $\{0, 1\}^*$. For example, numbers can be represented as binary strings using the binary encoding. (For readability, we identify numbers with their binary encodings in the examples below.)

Types of Computational Problems

A decision problem is a computational problem where the answer for every instance is either yes or no. An example of a decision problem is *primality testing*:

"Given a positive integer n, determine if n is prime."

A decision problem is typically represented as the set of all instances for which the answer is *yes*. For example, primality testing can be represented as the infinite set

$L = \{2, 3, 5, 7, 11, ...\}$

In a search problem, the answers can be arbitrary strings. For example, factoring is a search problem where the instances are (string representations of) positive integers and the solutions are (string representations of) collections of primes.

A search problem is represented as a relation over consisting of all the instance-solution pairs, called a *search relation*. For example, primality can be represented as the relation

R = {(4, 2), (6, 2), (6, 3), (8, 2), (8, 4), (9, 3), ...}

which consist of all pairs of numbers (n, p), where p is a nontrivial prime factor of n.

A counting problem asks for the number of solutions to a given search problem. For example, the counting problem associated with primality is "Given a positive integer n, count the number of nontrivial prime factors of n."

A counting problem can be represented by a function f from $\{0, 1\}^*$ to the nonnegative integers. For a search relation R, the counting problem associated to R is the function

$$f_R(\mathrm{x}) = |\{y : (x, y) \in R\}|.$$

An optimization problem asks for finding the "best possible" solution among the set of all possible solutions to a search problem. One example is the *maximum independent set* problem:

> *"Given a graph* G, *find an independent set of* G *of maximum size."*

Optimization problems can be represented by their search relations.

Promise Problems

In computational complexity theory, it is usually implicitly assumed that any string in $\{0, 1\}^*$represents an instance of the computational problem in question. However, sometimes not all strings $\{0, 1\}^*$ represent valid instances, and one specifies a proper subset of $\{0, 1\}^*$ as the set of "valid instances". Computational problems of this type are called promise problems.

The following is an example of a (decision) promise problem:

"Given a graph G, determine if every independent set in G has size at most 5, or G has an independent set of size at least 10."

Here, the valid instances are those graphs whose maximum independent set size is either at most 5 or at least 10.

Decision promise problems are usually represented as pairs of disjoint subsets (L_{yes}, L_{no}) of $\{0, 1\}^*$. The valid instances are those in $L_{yes} \cup L_{no}$. L_{yes} and L_{no} represent the instances whose answer is*yes* and *no*, respectively.

Promise problems play an important role in several areas of computational complexity, includinghardness of approximation, property testing, and interactive proof systems.

A computational problem can be viewed as an infinite collection of *instances* together with a *solution* for every instance. The input string for a computational problem is referred to as a problem instance, and should not be confused with the problem itself. In computational complexity theory, a problem refers to the abstract question to be solved. In contrast, an instance of this problem is a rather concrete utterance, which can serve as the input for a decision problem. For example, consider the problem of primality testing. The instance is a number (e.g. 15) and the solution is "yes" if the number is prime and "no" otherwise (in this case "no"). Alternatively, the instance is a particular input to the problem, and the solution is the output corresponding to the given input.

To further highlight the difference between a problem and an instance, consider the following instance of the decision version of the traveling salesman problem: Is there a route of at most 2000 kilometres in length passing through all of Germany's 15 largest cities? The answer to this particular problem instance is of little use for solving other instances of the problem, such as asking for a round trip through all sites in Milan whose total length is at most 10 km. For this reason, complexity theory addresses computational problems and not particular problem instances.

Representing Problem Instances

When considering computational problems, a problem instance is a string over an alphabet. Usually, the alphabet is taken to be the binary alphabet (i.e., the set {0,1}), and thus the strings are bitstrings. As in a real-world computer, mathematical objects other than bitstrings must be suitably encoded. For example, integers can be represented in binary notation, and graphs can be encoded directly via theiradjacency matrices, or by encoding their adjacency lists in binary.

Even though some proofs of complexity-theoretic theorems regularly assume some concrete choice of input encoding, one tries to keep the discussion abstract enough to be independent of the choice of encoding. This can be achieved by ensuring that different representations can be transformed into each other efficiently.

Decision Problem

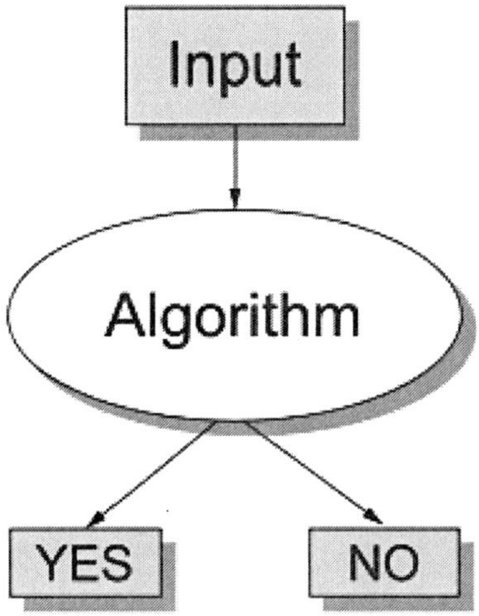

Figure: A decision problem has only two possible outputs, *yes* or *no* (or alternately 1 or 0) on any input.

In computability theory and computational complexity theory, adecision problem is a question in some formal system with a yes-or-no answer, depending on the values of some input parameters. For example, the problem "given two numbers $x$and y, does x evenly divide y?" is a decision problem. The answer can be either 'yes' or 'no', and depends upon the values of x and y.

Decision problems typically appear in mathematical questions of decidability, that is, the question of the existence of aneffective method to determine the existence of some object or its membership in a set; many of the important problems in mathematics are undecidable.

Decision problems are closely related to function problems, which can have answers that are more complex than a simple 'yes' or 'no'. A corresponding function problem is "given two numbers x and y, what is x divided by y?". They are also related to optimization problems, which are concerned with finding the *best* answer to a particular problem.

A method for solving a decision problem given in the form of an algorithm is called a decision procedure for that problem. A decision procedure for the decision problem "given two numbers x and$y$, does x evenly divide y?" would give the steps for determining whether x evenly divides y, given x and$y$. One such algorithm is long division, taught to many school children. If the remainder is zero the answer produced is 'yes', otherwise it is 'no'. A decision problem which can be solved by an algorithm, such as this example, is called decidable.

The field of computational complexity categorizes *decidable* decision problems by how difficult they are to solve. "Difficult", in this sense, is described in terms of the computational resources needed by the most efficient algorithm for a certain problem. The field of recursion theory, meanwhile, categorizes*undecidable* decision problems by Turing degree, which is a measure of the noncomputability inherent in any solution.

Research in computability theory has typically focused on decision problems. As explained in the section Equivalence with function problems below, there is no loss of generality.

Definition

A *decision problem* is any arbitrary yes-or-no question on an infinite set of inputs. Because of this, it is traditional to define the decision problem equivalently as: the set of inputs for which the problem returns *yes*.

These inputs can be natural numbers, but also other values of some other kind, such as strings of aformal language. Using some encoding, such as Gödel numberings, the strings can be encoded as natural numbers. Thus, a decision problem informally phrased in terms of a formal language is also equivalent to a set of natural numbers. To keep the formal definition simple, it is phrased in terms of subsets of the natural numbers.

Formally, a decision problem is a subset of the natural numbers. The corresponding informal problem is that of deciding whether a given number is in the set.

Examples

A classic example of a decidable decision problem is the set of prime numbers. It is possible to effectively decide whether a given natural number is prime by testing every possible nontrivial factor. Although much more efficient methods of primality testing are known, the existence of any effective method is enough to establish decidability.

Decidability

A decision problem A is called decidable or effectively solvable if A is a recursive set. A problem is called partially decidable, semidecidable, solvable, or provable if A is a recursively enumerable set. Partially decidable problems and any other problems that are not decidable are calledundecidable.

The halting problem is an important undecidable decision problem; for more examples, see list of undecidable problems.

Complete Problems

Decision problems can be ordered according to many-one reducibility and related feasible reductions such as Polynomial-time reductions. A decision problem P is said to be complete for a set of decision problems S if P is a member of S and every problem in S can be reduced to P. Complete decision problems are used in computational complexity to characterize complexity classes of decision problems. For example, the Boolean satisfiability problem is complete for the class NP of decision problems under polynomial-time reducibility.

History

The *Entscheidungsproblem*, German for "Decision-problem", is attributed to David Hilbert: "At [the] 1928 conference Hilbert made his questions quite precise. First, was mathematics *complete*... Second, was mathematics *consistent*... And thirdly, was mathematics *decidable*? By this he meant, did there exist a definite method which could, in principle be applied to any assertion, and which was guaranteed to produce a correct decision on whether that assertion was true" (Hodges, p. 91). Hilbert believed that "in mathematics there is no ignorabimus' (Hodges, p. 91ff) meaning 'we will not know'. See David Hilbert and Halting Problem for more.

Equivalence with Function Problems

A function problem consists of a partial function *f*; the informal "problem" is to compute the values of *f*on the inputs for which it is defined.

Every function problem can be turned into a decision problem; the decision problem is just the graph of the associated function. (The graph of a function *f* is the set of pairs (x,y) such that $f(x) = y$.) If this decision problem were effectively solvable then the function problem would be as well. This reduction does not respect computational complexity, however. For example, it is possible for the graph of a function to be decidable in polynomial time (in which case running time is computed as a function of the pair (x,y)) when the function is not computable in polynomial time (in which case running time is computed as a function of x alone). The function $f(x) = 2^x$ has this property.

Every decision problem can be converted into the function problem of computing the characteristic function of the set associated to the decision problem. If this function is computable then the associated decision problem is decidable. However, this reduction is more liberal than the standard reduction used in computational complexity (sometimes called polynomial-time many-one reduction); for example, the complexity of the characteristic functions of an NP-complete problem and its co-NP-complete complement is exactly the same even though the underlying decision problems may not be considered equivalent in some typical models of computation.

Practical Decision

Having practical decision procedures for classes of logical formulas is of considerable interest forprogram verification and circuit verification. Pure Boolean logical formulas are usually decided usingSAT-solving techniques based on the DPLL algorithm. Conjunctive formulas over linear real or rational arithmetic can be decided using the Simplex algorithm, formulas in linear integer arithmetic (Presburger arithmetic) can be decided using Cooper's algorithm or William Pugh's Omega test. Formulas with negations, conjunctions and disjunctions combine the difficulties of satisfiability testing with that of decision of conjunctions; they are generally decided nowadays using SMT-solvingtechnique, which combine SAT-solving with decision procedures for conjunctions and propagation techniques. Real polynomial arithmetic, also known as the theory of real closed fields, is decidable, for instance using the Cylindrical algebraic decomposition; unfortunately the complexity of that algorithm is excessive for most practical uses.

A leading scientific conference in this field is CAV. Decision problems are one of the central objects of study in computational complexity theory. A decision problem is a special type of computational problem whose answer is either*yes* or *no*, or alternately either 1 or 0. A decision problem can be viewed as a formal language, where the members of the language are instances whose answer is yes, and the non-members are those instances whose output is no. The objective is to decide, with the aid of an algorithm, whether a given input string is a member of the formal language under consideration. If the algorithm deciding this problem returns the answer *yes*, the algorithm is said to accept the input string, otherwise it is said to reject the input.

An example of a decision problem is the following. The input is an arbitrary graph. The problem consists in deciding whether the given graph is connected, or not. The formal language associated with this decision problem is then the set of all connected graphs—of course, to obtain a precise definition of this language, one has to decide how graphs are encoded as binary strings.

Function Problem

In computational complexity theory, a function problem is a computational problem where a single output (of a total function) is expected for every input, but the output is more complex than that of adecision problem, that is, it isn't just YES or NO. Notable examples include the travelling salesman problem, which asks for the route taken by the salesman, and the integer factorization problem, which asks for the list of factors.

Function problems are more awkward to study than decision problems because they don't have an obvious analogue in terms of languages, and because the notion of reduction between problems is more subtle as you have to transform the output as well as the input. Function problems can be sorted into complexity classes in the same way as decision problems. For example FP is the set of function problems which can be solved by a deterministic Turing machine in polynomial time, and FNP is the set of function problems which can be solved by a non-deterministic Turing machine in polynomial time.

For all function problems in which the solution is polynomially bounded, there is an analogous decision problem such that the function problem can be solved by polynomial-time Turing reduction to that decision problem. For example, the decision problem analogue to the travelling salesman problem is this:

Given a weighted directed graph and an integer K, is there a Hamilton path (or Hamilton cycle if the salesman must return home) with total weight less than or equal to K?

Given a solution to this problem, we can solve the travelling salesman problem as follows. Let n be the number of edges and w_i be the weight of edge i. First rescale and perturb the weights of the edges by assigning to edge i the new weight $w'_i = 2^{(n+1)}w_i + 2^i$. This doesn't change the shortest Hamilton path, but makes sure that it is unique. Now add the weights of all edges to get a total weight M. Find the weight of the shortest Hamilton path by binary search: is there a Hamilton path with weight$< M / 2$; is there a path with weight $< M / 4$ etc. Then having found the weight W of the shortest Hamilton path, determine which edges are in the path by asking for each edge i whether there is a Hamiltonian path with weight W for the graph modified so that edge i has weight $W + 1$ (if there is no such path in the modified graph, then edge i must be in the shortest path for the original graph).

This places the travelling salesman problem in the complexity class FP^{NP} (the class of function problems which can be solved in polynomial time on a deterministic Turing machine with an oracle for a problem in NP), and in fact it is complete for that cl

A function problem is a computational problem where a single output (of a total function) is expected for every input, but the output is more complex than that of a decision problem, that is, it isn't just yes or no. Notable examples include the traveling salesman problem and the integer factorization problem.

It is tempting to think that the notion of function problems is much richer than the notion of decision problems. However, this is not really the case, since function problems can be recast as decision problems. For example, the multiplication of two integers can be expressed as the set of triples (a, b, c) such that the relation $a \times b = c$ holds. Deciding whether a given triple is member of this set corresponds to solving the problem of multiplying two numbers.

Measuring the Size of an Instance

To measure the difficulty of solving a computational problem, one may wish to see how much time the best algorithm requires to solve the problem. However, the running time may, in general, depend on the instance. In particular, larger instances will require more time to solve. Thus the time required to solve a problem (or the space required, or any measure of complexity) is calculated as function of the size of the instance. This is usually taken to be the size of the input in bits. Complexity theory is interested in how algorithms scale with an increase in the input size. For instance, in the problem of finding whether a graph is connected, how much more time does it take to solve a problem for a graph with $2n$ vertices compared to the time taken for a graph with n vertices?

If the input size is n, the time taken can be expressed as a function of n. Since the time taken on different inputs of the same size can be different, the worst-case time complexity $T(n)$ is defined to be the maximum time taken over all inputs of size n. If $T(n)$ is a polynomial in n, then the algorithm is said to be a polynomial time algorithm. Cobham's thesis says that a problem can be solved with a feasible amount of resources if it admits a polynomial time algorithm.

Machine Models and Complexity Measures

Turing Machine

A Turing machine is a device that manipulates symbols on a strip of tape according to a table of rules. Despite its simplicity, a Turing machine can be adapted to simulate the logic of any computeralgorithm, and is particularly useful in explaining the functions of a CPUinside a computer.

The "Turing" machine was described by Alan Turing in 1936, who called it an "*a*(utomatic)-machine". The Turing machine is not intended as a practical computing technology, but rather as a hypothetical device representing a computing machine. Turing machines help computer scientists understand the limits of mechanical computation.

Turing gave a succinct definition of the experiment in his 1948 essay, "Intelligent Machinery". Referring to his 1936 publication, Turing wrote that the Turing machine, here called a Logical Computing Machine, consisted of: ...an infinite memory capacity obtained in the form of an infinite tape marked out into squares, on each of which a symbol could be printed. At any moment there is one symbol in the machine; it is called the scanned symbol. The machine can alter the scanned symbol and its behavior is in part determined by that symbol, but the symbols on the tape elsewhere do not affect the behaviour of the machine. However, the tape can be moved back and forth through the machine, this being one of the elementary operations of the machine. Any symbol on the tape may therefore eventually have an innings. (Turing 1948, p. 61)

A Turing machine that is able to simulate any other Turing machine is called a universal Turing machine (UTM, or simply a universal machine). A more mathematically-oriented definition with a similar "universal" nature was introduced by Alonzo Church, whose work on lambda calculus intertwined with Turing's in a formal theory of computation known as the Church–Turing thesis. The thesis states that Turing machines indeed capture the informal notion of effective method in logic and mathematics, and provide a precise definition of an algorithm or 'mechanical procedure'.

Studying their abstract properties yields many insights into computer science and complexity theory.

Informal Description

The Turing machine mathematically models a machine that mechanically operates on a tape. On this tape are symbols which the machine can read and write, one at a time, using a tape head. Operation is fully determined by a finite set of elementary instructions such as "in state 42, if the symbol seen is 0, write a 1; if the symbol seen is 1, shift to the right, and change into state 17; in state 17, if the symbol seen is 0, write a 1 and change to state 6;" etc. In the original article ("On computable numbers, with an application to the Entscheidungsproblem", see also references below), Turing imagines not a mechanism, but a person whom he calls the "computer", who executes these deterministic mechanical rules slavishly (or as Turing puts it, "in a desultory manner").

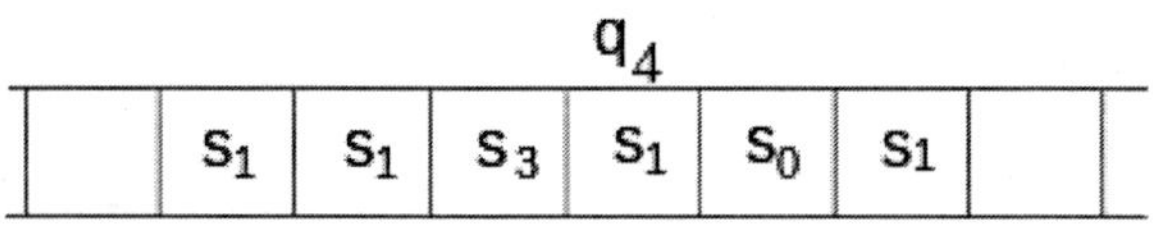

Figure: The head is always over a particular square of the tape; only a finite stretch of squares is given. The instruction to be performed (q_4) is shown over the scanned square. (Drawing after Kleene (1952) p.375.)

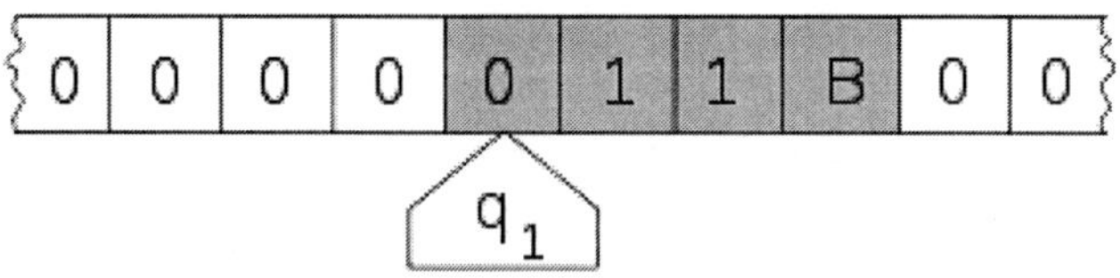

Figure: Here, the internal state (q_1) is shown inside the head, and the illustration describes the tape as being infinite and pre-filled with "0", the symbol serving as blank. The system's full state (its *configuration*) consists of the internal state, the contents of the shaded squares including the blank scanned by the head ("11B"), and the position of the head. (Drawing after Minsky (1967) p. 121).

More precisely, a Turing machine consists of:

1. A tape which is divided into cells, one next to the other. Each cell contains a symbol from some finite alphabet. The alphabet contains a special *blank*symbol (here written as ‘B’) and one or more other symbols. The tape is assumed to be arbitrarily extendable to the left and to the right, i.e., the Turing machine is always supplied with as much tape as it needs for its computation. Cells that have not been written to before are assumed to be filled with the blank symbol. In some models the tape has a left end marked with a special symbol; the tape extends or is indefinitely extensible to the right.
2. A head that can read and write symbols on the tape and move the tape left and right one (and only one) cell at a time. In some models the head moves and the tape is stationary.
3. A finite table (occasionally called an action table or transition function) of instructions (usually quintuples [5-tuples] : $q_i a_j$’!$q_{i1} a_{j1} d_k$, but sometimes 4-tuples) that, given the *state*(q_i) the machine is currently in *and* the *symbol*(a_j) it is reading on the tape (symbol currently under the head) tells the machine to do the following in sequence (for the 5-tuple models):
   - Either erase or write a symbol (instead of a_j, write a_{j1}), *and then*
   - Move the head (which is described by d_k and can have values: ‘L’ for one step left *or* ’R’ for one step right *or* ’N’ for staying in the same place), *and then*
   - Assume the same or a *new state* as prescribed (go to state q_{i1}).

   In the 4-tuple models, erase or write a symbol (a_{j1}) and move the head left or right (d_k) are specified as separate instructions. Specifically, the table tells the machine to (ia) erase or write a symbol *or* (ib) move the head left or right, *and then* (ii) assume the same or a new state as prescribed, but not both actions (ia) and (ib) in the same instruction. In some models, if there is no entry in the table for the current combination of symbol and state then the machine will halt; other models require all entries to be filled.
4. A state register that stores the state of the Turing machine, one of finitely many. There is one special *start state* with which the state register is initialized. These states, writes Turing,

replace the "state of mind" a person performing computations would ordinarily be in.

Note that every part of the machine—its state and symbol-collections—and its actions—printing, erasing and tape motion—is *finite*, *discrete* and *distinguishable*; it is the potentially unlimited amount of tape that gives it an unbounded amount of storage space.

Examples of Turing Machines

To see examples of the following models, see Turing machine examples:

1. Turing's very first machine
2. Copy routine
3. 3-state busy beaver.

Formal Definition

Hopcroft and Ullman (1979, p. 148) formally define a (one-tape) Turing machine as a 7-tuple $M = \langle Q,\Gamma,b,\Sigma,\delta,q_0,F\rangle$ where

- Q is a finite, non-empty set of *states*
- Ã is a finite, non-empty set of the *tape alphabet/symbols*
- $b \in \Gamma$ is the *blank symbol* (the only symbol allowed to occur on the tape infinitely often at any step during the computation)
- $\Sigma \subseteq \Gamma \setminus \{b\}$ is the set of *input symbols*
- $q_0 \in Q$ is the *initial state*
- $F \subseteq Q$ is the set of *final* or *accepting states.*
- $\delta : Q \setminus F \times \Gamma \rightarrow Q \times \Gamma \times \{L,R\}$ is a partial function called the *transition function*, where L is left shift, R is right shift. (A relatively uncommon variant allows "no shift", say N, as a third element of the latter set.)

Anything that operates according to these specifications is a Turing machine.

The 7-tuple for the 3-state busy beaver looks like this (see more about this busy beaver at Turing machine examples):

Q = { A, B, C, HALT }

Ã = { 0, 1 }

b = 0 = "blank"

Σ = { 1 }

δ = see state-table below

q_0 = A = initial state

F = the one element set of final states {HALT}

Initially all tape cells are marked with 0.

State table for 3 state, 2 symbol busy beaver									
Tape symbol	Current state A			Current state B			Current state C		
	Write symbol	Move tape	Next state	Write symbol	Move tape	Next state	Write symbol	Move tape	Next state
0	1	R	B	1	L	A	1	L	B
1	1	L	C	1	R	B	1	R	HALT

Additional Details Required to Visualize or Implement Turing Machines

In the words of van Emde Boas (1990), p. 6: "The set-theoretical object [his formal seven-tuple description similar to the above] provides only partial information on how the machine will behave and what its computations will look like."

For instance:

- There will need to be some decision on what the symbols actually look like, and a failproof way of reading and writing symbols indefinitely.
- The shift left and shift right operations may shift the tape head across the tape, but when actually building a Turing machine it is more practical to make the tape slide back and forth under the head instead.
- The tape can be finite, and automatically extended with blanks as needed (which is closest to the mathematical definition), but it is more common to think of it as stretching infinitely at both ends and being pre-filled with blanks except on the explicitly given finite fragment the tape head is on. (This is, of course, not implementable in practice.) The tape *cannot* be fixed in length, since that would not correspond to the given definition and would seriously limit the range of computations the machine can perform to those of a linear bounded automaton.

The "State"

The word "state" used in context of Turing machines can be a source of confusion, as it can mean two things. Most commentators after Turing have used "state" to mean the name/designator of the current instruction to be performed—i.e. the contents of the state register. But Turing (1936) made a strong distinction between a record of what he called the machine's "m-configuration", (its internal state) and the machine's (or person's) "state of progress" through the computation - the current state of the total system. What Turing called "the state formula" includes both the current instruction and *all* the symbols on the tape:

> Thus the state of progress of the computation at any stage is completely determined by the note of instructions and the symbols on the tape. That is, the state of the systemmay be described by a single expression (sequence of symbols) consisting of the symbols on the tape followed by Δ (which we suppose not to appear elsewhere) and then by the note of instructions. This expression is called the 'state formula'.
>
> —*Undecidable*, p.139–140, emphasis added

Earlier in his paper Turing carried this even further: he gives an example where he places a symbol of the current "m-configuration"—the instruction's label—beneath the scanned square, together with all the symbols on the tape (*Undecidable*, p. 121); this he calls "the *complete configuration*" (*Undecidable*, p. 118). To print the "complete configuration" on one line he places the state-label/m-configuration to the *left* of the scanned symbol.

A variant of this is seen in Kleene (1952) where Kleene shows how to write the Gödel number of a machine's "situation": he places the "m-configuration" symbol q_4 over the scanned square in roughly the center of the 6 non-blank squares on the tape (see the Turing-tape figure in this article) and puts it to the *right* of the scanned square. But Kleene refers to "q_4" itself as "the machine state" (Kleene, p. 374-375). Hopcroft and Ullman call this composite the "instantaneous description" and follow the Turing convention of putting the "current state" (instruction-label, m-configuration) to the *left* of the scanned symbol (p. 149).

Example: total state of 3-state 2-symbol busy beaver after 3 "moves" (taken from example "run" in the figure below):

1A1

This means: after three moves the tape has ... 000110000 ... on it, the head is scanning the right-most 1, and the state is A. Blanks (in this case represented by "0"s) can be part of the total state as shown here: B01 ; the tape has a single 1 on it, but the head is scanning the 0 ("blank") to its left and the state is B.

"State" in the context of Turing machines should be clarified as to which is being described: (*i*) the current instruction, or (*ii*) the list of symbols on the tape together with the current instruction, or (*iii*) the list of symbols on the tape together with the current instruction placed to the left of the scanned symbol or to the right of the scanned symbol.

Turing's biographer Andrew Hodges (1983: 107) has noted and discussed this confusion.

Models Equivalent to the Turing Machine Model

Many machines that might be thought to have more computational capability than a simple universal Turing machine can be shown to have no more power (Hopcroft and Ullman p. 159, cf Minsky (1967)). They might compute faster, perhaps, or use less memory, or their instruction set might be smaller, but they cannot compute more powerfully (i.e. more mathematical functions). (Recall that the Church–Turing thesis *hypothesizes* this to be true for any kind of machine: that anything that can be "computed" can be computed by some Turing machine.)

A Turing machine is equivalent to a pushdown automaton that has been made more flexible and concise by relaxing the last-in-first-out requirement of its stack.

At the other extreme, some very simple models turn out to be Turing-equivalent, i.e. to have the same computational power as the Turing machine model.

Common equivalent models are the multi-tape Turing machine, multi-track Turing machine, machines with input and output, and the *non-deterministic* Turing machine (NDTM) as opposed to the*deterministic* Turing machine (DTM) for which the action table has at most one entry for each combination of symbol and state.

Read-only, right-moving Turing machines are equivalent to NDFAs (as well as DFAs by conversion using the NDFA to DFA conversion algorithm).

For practical and didactical intentions the equivalent register machine can be used as a usualassembly programming language.

Choice c-machines, Oracle o-machines

Early in his paper (1936) Turing makes a distinction between an "automatic machine"—its "motion ... completely determined by the configuration" and a "choice machine":

...whose motion is only partially determined by the configuration ... When such a machine reaches one of these ambiguous configurations, it cannot go on until some arbitrary choice has been made by an external operator. This would be the case if we were using machines to deal with axiomatic systems. —*Undecidable*, p. 118

Turing (1936) does not elaborate further except in a footnote in which he describes how to use an a-machine to "find all the provable formulae of the [Hilbert] calculus" rather than use a choice machine. He "suppose[s] that the choices are always between two possibilities 0 and 1. Each proof will then be determined by a sequence of choices $i_1, i_2, ..., i_n$ ($i_1 = 0$ or 1, $i_2 = 0$ or 1, ..., $i_n = 0$ or 1), and hence the number $2^n + i_1 2^{n-1} + i_2 2^{n-2} + ... + i_n$ completely determines the proof. The automatic machine carries out successively proof 1, proof 2, proof 3, ..." (Footnote ‡, *Undecidable*, p. 138)

This is indeed the technique by which a deterministic (i.e. a-) Turing machine can be used to mimic the action of a nondeterministic Turing machine; Turing solved the matter in a footnote and appears to dismiss it from further consideration.

An oracle machine or o-machine is a Turing a-machine that pauses its computation at state "o" while, to complete its calculation, it "awaits the decision" of "the oracle"—an unspecified entity "apart from saying that it cannot be a machine" (Turing (1939), Undecidable p. 166–168). The concept is now actively used by mathematicians.

Universal Turing Machines

As Turing wrote in *Undecidable*, p. 128 (italics added):

It is possible to invent a *single machine* which can be used to compute *any* computable sequence. If this machine U is supplied with the tape on the beginning of which is written the string of quintuples separated by semicolons of some computing machine M, then Uwill compute the same sequence as M. This finding is now taken for granted, but at the time (1936) it was considered astonishing. The model of computation that Turing called his "universal machine"—"U" for short—is considered by some (cf Davis (2000)) to have been the fundamental theoretical breakthrough that led to the notion of theStored-program computer.

Turing's paper ... contains, in essence, the invention of the modern computer and some of the programming techniques that accompanied it. —Minsky (1967), p. 104

In terms of computational complexity, a multi-tape universal Turing machine need only be slower bylogarithmic factor compared to the machines it simulates. This result was obtained in 1966 by F. C. Hennie and R. E. Stearns. (Arora and Barak, 2009, theorem 1.9)

Comparison with Real Machines

It is often said that Turing machines, unlike simpler automata, are as powerful as real machines, and are able to execute any operation that a real program can.

What is missed in this statement is that, because a real machine can only be in finitely many *configurations*, in fact this "real machine" is nothing but a linear bounded automaton.

On the other hand, Turing machines are equivalent to machines that have an unlimited amount of storage space for their computations. In fact, Turing machines are not intended to model computers, but rather they are intended to model computation itself; historically, computers, which compute only on their (fixed) internal storage, were developed only later.

There are a number of ways to explain why Turing machines are useful models of real computers:

1. Anything a real computer can compute, a Turing machine can also compute. For example: "A Turing machine can simulate any type of subroutine found in programming languages, including recursive procedures and any of the known parameter-passing mechanisms" (Hopcroft and Ullman p. 157). A large enough FSA can also model any real computer, disregarding IO. Thus, a statement about the limitations of Turing machines will also apply to real computers.
2. The difference lies only with the ability of a Turing machine to manipulate an unbounded amount of data. However, given a finite amount of time, a Turing machine (like a real machine) can only manipulate a finite amount of data.

3. Like a Turing machine, a real machine can have its storage space enlarged as needed, by acquiring more disks or other storage media. If the supply of these runs short, the Turing machine may become less useful as a model. But the fact is that neither Turing machines nor real machines need astronomical amounts of storage space in order to perform useful computation. The processing time required is usually much more of a problem.
4. Descriptions of real machine programs using simpler abstract models are often much more complex than descriptions using Turing machines. For example, a Turing machine describing an algorithm may have a few hundred states, while the equivalent deterministic finite automaton on a given real machine has quadrillions. This makes the DFA representation infeasible to analyze.
5. Turing machines describe algorithms independent of how much memory they use. There is a limit to the memory possessed by any current machine, but this limit can rise arbitrarily in time. Turing machines allow us to make statements about algorithms which will (theoretically) hold forever, regardless of advances in *conventional* computing machine architecture.
6. Turing machines simplify the statement of algorithms. Algorithms running on Turing-equivalent abstract machines are usually more general than their counterparts running on real machines, because they have arbitrary-precision data types available and never have to deal with unexpected conditions (including, but not limited to, running out of memory).

One way in which Turing machines are a poor model for programs is that many real programs, such asoperating systems and word processors, are written to receive unbounded input over time, and therefore do not halt. Turing machines do not model such ongoing computation well (but can still model portions of it, such as individual procedures).

Limitations of Turing Machines

Computational Complexity Theory

A limitation of Turing machines is that they do not model the strengths of a particular arrangement well. For instance, modern stored-program computers are actually instances of a more specific form ofabstract machine known as the random access stored program machine or RASP machine model. Like the Universal Turing machine the RASP stores its "program" in "memory" external to its finite-state machine's "instructions". Unlike the universal Turing machine, the RASP has an infinite number of distinguishable, numbered but unbounded "registers"—memory "cells" that can contain any integer (cf. Elgot and Robinson (1964), Hartmanis (1971), and in particular Cook-Rechow (1973); references atrandom access machine). The RASP's finite-state machine is equipped with the capability for indirect addressing (e.g. the contents of one register can be used as an address to specify another register); thus the RASP's "program" can address any register in the register-sequence. The upshot of this distinction is that there are computational optimizations that can be performed based on the memory indices, which are not possible in a general Turing machine; thus when Turing machines are used as the basis for bounding running times, a 'false lower bound' can be proven on certain algorithms' running times (due to the false simplifying assumption of a Turing machine). An example of this is binary search, an algorithm that can be shown to perform more quickly when using the RASP model of computation rather than the Turing machine model.

Concurrency

Another limitation of Turing machines is that they do not model concurrency well. For example, there is a bound on the size of integer that can be computed by an always-halting nondeterministic Turing machine starting on a blank tape. (See article on unbounded nondeterminism.) By contrast, there are always-halting concurrent systems with no inputs that can compute an integer of unbounded size. (A process can be created with local storage that is initialized with a count of 0 that concurrently sends itself both a stop and a go message. When it receives a go message, it increments its count by 1 and sends itself a go message. When it receives a stop message, it stops with an unbounded number in its local storage.)

History

They were described in 1936 by Alan Turing.

Historical Background: Computational Machinery

Robin Gandy (1919–1995)—a student of Alan Turing (1912–1954) and his life-long friend—traces the lineage of the notion of "calculating machine" back to Babbage (circa 1834) and actually proposes "Babbage's Thesis":

That the whole of development and operations of analysis are now capable of being executed by machinery. —(italics in Babbage as cited by Gandy, p. 54)

Gandy's analysis of Babbage's Analytical Engine describes the following five operations (cf p. 52–53):

1. The arithmetic functions +, ", × where " indicates "proper" subtraction x – y = 0 if y e– x
2. Any sequence of operations is an operation
3. Iteration of an operation (repeating n times an operation P)
4. Conditional iteration (repeating n times an operation P conditional on the "success" of test T)
5. Conditional transfer (i.e. conditional "goto").

Gandy states that "the functions which can be calculated by (1), (2), and (4) are precisely those which are Turing computable." (p. 53). He cites other proposals for "universal calculating machines" included those of Percy Ludgate (1909), Leonardo Torres y Quevedo (1914), Maurice d'Ocagne (1922), Louis Couffignal (1933), Vannevar Bush (1936), Howard Aiken (1937). However:

... the emphasis is on programming a fixed iterable sequence of arithmetical operations. The fundamental importance of conditional iteration and conditional transfer for a general theory of calculating machines is not recognized ... —Gandy p. 55

The Entscheidungsproblem (the "decision problem"): Hilbert's tenth question of 1900

With regards to Hilbert's problems posed by the famous mathematician David Hilbert in 1900, an aspect of problem #10 had been floating about for almost 30 years before it was framed precisely. Hilbert's original expression for #10 is as follows:

10. Determination of the solvability of a Diophantine equation. Given aDiophantine equation with any number of unknown quantities and with rational integral coefficients: To devise a process according to which it can be determined in a finite number of operations whether the equation is solvable in rational integers. The Entscheidungsproblem [decision problem for first-order logic] is solved when we know a procedure that allows for any given logical expression to decide by

finitely many operations its validity or satisfiability ... The Entscheidungsproblem must be considered the main problem of mathematical logic.

—quoted, with this translation and the original German, in Dershowitz and Gurevich, 2008

By 1922, this notion of "Entscheidungsproblem" had developed a bit, and H. Behmann stated that

... most general form of the Entscheidungsproblem [is] as follows:

A quite definite generally applicable prescription is required which will allow one to decide in a finite number of steps the truth or falsity of a given purely logical assertion ... —Gandy p. 57, quoting Behmann

Behmann remarks that ... the general problem is equivalent to the problem of deciding which mathematical propositions are true. —*ibid.*

If one were able to solve the Entscheidungsproblem then one would have a "procedure for solving many (or even all) mathematical problems". —*ibid.*, p. 92

By the 1928 international congress of mathematicians Hilbert "made his questions quite precise. First, was mathematics *complete* ... Second, was mathematics *consistent* ... And thirdly, was mathematics*decidable*?" (Hodges p. 91, Hawking p. 1121). The first two questions were answered in 1930 by Kurt Gödel at the very same meeting where Hilbert delivered his retirement speech (much to the chagrin of Hilbert); the third—the Entscheidungsproblem—had to wait until the mid-1930s.

The problem was that an answer first required a precise definition of "*definite general applicable prescription*", which Princeton professor Alonzo Church would come to call "effective calculability", and in 1928 no such definition existed. But over the next 6–7 years Emil Post developed his definition of a worker moving from room to room writing and erasing marks per a list of instructions (Post 1936), as did Church and his two students Stephen Kleene and J. B. Rosser by use of Church's lambda-calculus and Gödel's recursion theory (1934). Church's paper (published 15 April 1936) showed that the Entscheidungsproblem was indeed "undecidable" and beat Turing to the punch by almost a year (Turing's paper submitted 28 May 1936, published January 1937). In the meantime, Emil Post submitted a brief paper in the fall of 1936, so Turing at least had priority over Post. While Church refereed Turing's paper, Turing had time to study Church's paper and add an Appendix where he sketched a proof that Church's lambda-calculus and his machines would compute the same functions.

But what Church had done was something rather different, and in a certain sense weaker. ... the Turing construction was more direct, and provided an argument from first principles, closing the gap in Church's demonstration. —Hodges p. 112

And Post had only proposed a definition of calculability and criticized Church's "definition", but had proved nothing.

Alan Turing's a- (Automatic-) Machine

In the spring of 1935 Turing as a young Master's student at King's College Cambridge, UK, took on the challenge; he had been stimulated by the lectures of the logician M. H. A. Newman "and learned from them of Gödel's work and the Entscheidungsproblem ... Newman used the word 'mechanical' ... In his obituary of Turing 1955 Newman writes:

To the question 'what is a "mechanical" process?' Turing returned the characteristic answer 'Something that can be done by a machine' and he embarked on the highly congenial task of analysing the general notion of a computing machine. —Gandy, p. 74

Gandy states that: I suppose, but do not know, that Turing, right from the start of his work, had as his goal a proof of the undecidability of the Entscheidungsproblem. He told me that the 'main idea' of the paper came to him when he was lying in Grantchester meadows in the summer of 1935. The 'main idea' might have either been his analysis of computation or his realization that there was a universal machine, and so a diagonal argument to prove unsolvability.—*ibid.*, p. 76

While Gandy believed that Newman's statement above is "misleading", this opinion is not shared by all. Turing had a life-long interest in machines: "Alan had dreamt of inventing typewriters as a boy; [his mother] Mrs. Turing had a typewriter; and he could well have begun by asking himself what was meant by calling a typewriter 'mechanical'" (Hodges p. 96).

While at Princeton pursuing his PhD, Turing built a Boolean-logic multiplier (see below). His PhD thesis, titled "Systems of Logic Based on Ordinals", contains the following definition of "a computable function":

It was stated above that 'a function is effectively calculable if its values can be found by some purely mechanical process'. We may take this statement literally, understanding by a purely mechanical process one which could be carried out by a machine. It is possible to give a mathematical description, in a certain normal form, of the structures of these machines. The development of these ideas leads to the author's definition of a computable function, and to an identification of computability with effective calculability. It is not difficult, though somewhat laborious, to prove that these three definitions [the 3rd is the λ-calculus] are equivalent. —Turing (1939) in *The Undecidable*, p. 160

When Turing returned to the UK he ultimately became jointly responsible for breaking the German secret codes created by encryption machines called "The Enigma"; he also became involved in the design of the ACE (Automatic Computing Engine), "[Turing's] ACE proposal was effectively self-contained, and its roots lay not in the EDVAC [the USA's initiative], but in his own universal machine" (Hodges p. 318). Arguments still continue concerning the origin and nature of what has been named by Kleene (1952) Turing's Thesis. But what Turing *did prove* with his computational-machine model appears in his paper *On Computable Numbers, With an Application to the Entscheidungsproblem*(1937):

[that] the Hilbert Entscheidungsproblem can have no solution ... I propose, therefore to show that there can be no general process for determining whether a given formula U of the functional calculus K is provable, i.e. that there can be no machine which, supplied with any one U of these formulae, will eventually say whether U is provable. —from Turing's paper as reprinted in *The Undecidable*, p. 145

Turing's example (his second proof): If one is to ask for a general procedure to tell us: "Does this machine ever print 0", the question is "undecidable".

1937–1970: The "Digital Computer", the Birth of "Computer Science"

In 1937, while at Princeton working on his PhD thesis, Turing built a digital (Boolean-logic) multiplier from scratch, making his own electromechanical relays (Hodges p. 138). "Alan's task was to embody the logical design of a Turing machine in a network of relay-operated switches ..." (Hodges p. 138). While Turing might have been just curious and experimenting, quite-earnest work in the same direction was going in Germany (Konrad Zuse (1938)), and in the United States (Howard Aiken) and George Stibitz (1937); the fruits of their labors were used by the Axis and Allied military in World War II (cf Hodges p. 298–299). In the early to mid-1950s Hao Wang and Marvin Minsky reduced the Turing machine to a simpler form (a precursor to the Post-Turing machine of Martin Davis); simultaneously European researchers were reducing the new-fangled electronic computer to a computer-like theoretical object equivalent to what was now being called a "Turing machine". In the late 1950s and early 1960s, the coincidentally-parallel developments of Melzak and Lambek (1961), Minsky (1961), and Shepherdson and Sturgis (1961) carried the European work further and reduced the Turing machine to a more friendly, computer-like abstract model called the counter machine; Elgot and Robinson (1964), Hartmanis (1971), Cook and Reckhow (1973) carried this work even further with theregister machine and random access machine models—but basically all are just multi-tape Turing machines with an arithmetic-like instruction set.

1970–Present: the Turing Machine as a Model of Computation

Today the counter, register and random-access machines and their sire the Turing machine continue to be the models of choice for theorists investigating questions in the theory of computation. In particular, computational complexity theory makes use of the Turing machine:

Depending on the objects one likes to manipulate in the computations (numbers like nonnegative integers or alphanumeric strings), two models have obtained a dominant position in machine-based complexity theory: *the off-line multitape Turing machine...*, which represents the standard model for string-oriented computation, and the *random access machine (RAM)* as introduced by Cook and Reckhow ..., which models the idealized Von Neumann style computer. —van Emde Boas 1990:4

Only in the related area of analysis of algorithms this role is taken over by the RAM model. —van Emde Boas 1990:16

Kantorovitz (2005), was the first to show the most simple obvious representation of Turing machines published academically which unifies Turing machines with mathematical analysis and analog computers.

A Turing machine is a mathematical model of a general computing machine. It is a theoretical device that manipulates symbols contained on a strip of tape. Turing machines are not intended as a practical computing technology, but rather as a thought experiment representing a computing machine. It is believed that if a problem can be solved by an algorithm, there exists a Turing machine that solves the problem. Indeed, this is the statement of the Church–Turing thesis. Furthermore, it is known that everything that can be computed on other models of computation known to us today, such as a RAM machine, Conway's Game of Life, cellular automata or any programming language can be computed on a Turing machine. Since Turing machines are easy to analyze mathematically, and are believed to be as powerful as any other model of computation, the Turing machine is the most commonly used model in complexity theory. Many types of Turing machines are used to define complexity classes, such as deterministic Turing machines, probabilistic Turing machines, non-deterministic Turing machines, quantum Turing machines, symmetric Turing machines and alternating Turing machines. They are all equally powerful in principle, but when resources (such as time or space) are bounded, some of these may be more powerful than others.

A deterministic Turing machine is the most basic Turing machine, which uses a fixed set of rules to determine its future actions. A probabilistic Turing machine is a deterministic Turing machine with an extra supply of random bits. The ability to make probabilistic decisions often helps algorithms solve problems more efficiently. Algorithms that use random bits are called randomized algorithms. A non-deterministic Turing machine is a deterministic Turing machine with an added feature of non-determinism, which allows a Turing machine to have multiple possible future actions from a given state. One way to view non-determinism is that the Turing machine branches into many possible computational paths at each step, and if it solves the problem in any of these branches, it is said to have solved the problem. Clearly, this model is not meant to be a physically realizable model, it is just a theoretically interesting abstract machine that gives rise to particularly interesting complexity classes. For examples, see nondeterministic algorithm.

Turing Machine Equivalents

Machines Equivalent to the Turing Machine Model

Turing equivalence: Many machines that might be thought to have more computational capability than a simple universal Turing machine can be shown to have no more power (Hopcroft and Ullman

p. 159, cf Minsky). They might compute faster, perhaps, or use less memory, or their instruction set might be smaller, but they cannot compute more powerfully (i.e. more mathematical functions). (Recall that the Church-Turing thesis*hypothesizes* this to be true: that anything that can be "computed" can be computed by some Turing machine.)

While none of the following models have been shown to have more power than the single-tape, one-way infinite, multi-symbol Turing-machine model, their authors defined and used them to investigate questions and solve problems more easily than they could have if they had stayed with Turing's a-machine model.

The Sequential-machine Models: All of the following are called "sequential machine models" to distinguish them from "parallel machine models" (van Emde Boas (1990) p. 18).

Tape-based Turing Machines

Turing's a-machine model: Turing's (1936) a-machine (his name) was left-ended, right-end-infinite. He provided symbols YY to mark the left end. Any of finite number of tape symbols were permitted. The instructions (if a universal machine), and the "input" and "out" were written on only on "F-squares", and markers were to appear on "E-squares". In essence he divided his machine into two tapes that always moved together. The instructions appeared in a tabular form called "5-tuples" and were not executed sequentially.

Single-tape Machines with Restricted Symbols and/or Restricted Instructions

The following models are single tape Turing machines but restricted with (i) restricted tape symbols { mark, blank }, and/or (ii) sequential, computer-like instructions, and/or (iii) machine-actions fully atomized.

Post's "Formulation 1" Model of Computation

Emil Post (1936) in an independent description of a computational process, reduced the symbols allowed to the equivalent binary set of marks on the tape { "mark", "blank"=not_mark }. He changed the notion of "tape" from 1-way infinite to the right to an infinite set of rooms each with a sheet of paper in both directions. He atomized the Turing 5-tuples into 4-tuples—motion instructions separate from print/erase instructions. Although his (1936) model is ambiguous about this, Post's (1947) model did not require sequential instruction execution.

His extremely simple model can emulate any Turing machine, and although his 1936 *Formulation 1*does not use the word "program" or "machine", it is effectively a formulation of a very primitive programmable computer and associated programming language, with the boxes acting as an unbounded bitstring memory, and the set of instructions constituting a program.

Wang Machines

In an influential paper, Hao Wang (1954, 1957) reduced Post's "formulation 1" to machines that still use a two-way infinite binary tape, but whose instructions are simpler — being the "atomic" components of Post's instructions — and are by default executed sequentially (like a "computer program"). His stated principal purpose was to offer, as an alternative to Turing's theory, one that "is more economical in the basic operations". His results were "program formulations" of a variety of such machines, including the 5-instruction Wang W-machine with the instruction-set

{SHIFT-LEFT, SHIFT-RIGHT, MARK-SQUARE, ERASE-SQUARE, JUMP-IF-SQUARE-MARKED-to xxx}

and his most-severely reduced 4-instruction Wang B-machine ("B" for "basic") with the instruction-set

{SHIFT-LEFT, SHIFT-RIGHT, MARK-SQUARE, JUMP-IF-SQUARE-MARKED-to xxx}

which has not even an ERASE-SQUARE instruction.

Many authors later introduced variants of the machines discussed by Wang:

Minsky (1961) evolved Wang's notion with his version of the (multi-tape) "counter machine" model that allowed SHIFT-LEFT and SHIFT-RIGHT motion of the separate heads but no printing at all. In this case the tapes would be left-ended, each end marked with a single "mark" to indicate the end. He was able to reduce this to a single tape, but at the expense of introducing multi-tape-square motion equivalent to multiplication and division rather than the much simpler { SHIFT-LEFT = DECREMENT, SHIFT-RIGHT = INCREMENT }.

Davis, adding an explicit HALT instruction to one of the machines discussed by Wang, used a model with the instruction-set

{ SHIFT-LEFT, SHIFT-RIGHT, ERASE, MARK, JUMP-IF-SQUARE-MARKED-to xxx, JUMP-to xxx, HALT } and also considered versions with tape-alphabets of size larger than 2.

Böhm's Theoretical Machine Language P"

In keeping with Wang's project to seek a Turing-equivalent theory "economical in the basic operations", and wishing to avoid unconditional jumps, a notable theoretical language is the 4-instruction language P" introduced by Corrado Böhm in 1964 — the first "GOTO-less" imperative "structured programming" language to be proved Turing-complete.

Multi-tape Turing Machines

In practical analysis, various types of multi-tape Turing machines are often used. Multi-tape machines are similar to single-tape machines, but there is some constant k number of independent tapes.

The TABLE has full independent control over all the heads, any of all of which move and print/erase their own tapes (cf Aho-Hopcroft-Ullman 1974 p. 26). Most models have tapes with left ends, right ends unbounded.

This model intuitively seems much more powerful than the single-tape model, but any multi-tape machine, no matter how large the k, can be simulated by a single-tape machine using only quadratically more computation time (Papadimitriou 1994, Thrm 2.1). Thus, multi-tape machines cannot calculate any more functions than single-tape machines, and none of the robust complexity classes (such as polynomial time) are affected by a change between single-tape and multi-tape machines.

Two-stack Turing Machine

Two-stack Turing machines have a read-only input and two storage tapes. If a head moves left on either tape a blank is printed on that tape, but one symbol from a "library" can be printed.

Formal Definition: Multi-tape Turing Machine

A k-tape Turing machine can be described as a 6-tuple $M = \langle Q, \Gamma, s, b, F, \delta \rangle$ where

- Q is a finite set of states
- Ã is a finite set of the tape alphabet
- $s \in Q$ is the initial state
- $b \in \Gamma$ is the blank symbol

- $F \subseteq Q$ is the set of final or accepting states
- $\delta : Q \times \Gamma^k \rightarrow Q \times (\Gamma \times \{L, R, S\})^k$ is a partial function called the transition function, where L is left shift, R is right shift, S is no shift.

Deterministic and non-deterministic Turing machines

If the action table has at most one entry for each combination of symbol and state then the machine is a "deterministic Turing machine" (DTM). If the action table contains multiple entries for a combination of symbol and state then the machine is a "non-deterministic Turing machine" (NDTM). The two are computationally equivalent, that is, it is possible to turn any NDTM into a DTM (and *vice versa*).

Oblivious Turing Machines

An oblivious Turing machine is a Turing machine where movement of the various heads are fixed functions of time, independent of the input. In other words, there is a predetermined sequence in which the various tapes are scanned, advanced, and written to. Pippenger and Fischer (1979) showed that any computation that can be performed by a multi-tape Turing machine in *n* steps can be performed by an oblivious two-tape Turing machine in *O(n log n)* steps.

Register Machine Models

van Emde Boas (1990) includes all machines of this type in one category (group, class, collection) — "the register machine". However, historically the literature has also called the most primitive member of this group i.e. "the counter machine" — "the register machine". And the most primitive embodiment of a "counter machine" is sometimes called the "Minsky machine".

The "Counter Machine", Also Called a "Register Machine" Model

The primitive model register machine is, in effect, a multitape 2-symbol Post-Turing machine with its behavior restricted so its tapes act like simple "counters".

By the time of Melzak, Lambek, and Minsky (all 1961) the notion of a "computer program" produced a different type of simple machine with many left-ended tapes cut from a Post-Turing tape. In all cases the models permit only two tape symbols { mark, blank }.

Some versions represent the positive integers as only a strings/ stack of marks allowed in a "register" (i.e. left-ended tape), and a blank tape represented by the count "0". Minsky (1961) eliminated the PRINT instruction at the expense of providing his model with a mandatory single mark at the left-end of each tape.

In this model the single-ended tapes-as-registers are thought of as "counters", their instructions restricted to only two (or three if the TEST/DECREMENT instruction is atomized). Two common instruction sets are the following:

(1): { INC (r), DEC (r), JZ (r,z) }, i.e.

{ INCrement contents of register #r; DECrement contents of register #r; IF contents of #r=Zero THEN Jump-to Instruction #z}

(2): { CLR (r); INC (r); JE (r_i, r_j, z) }, i.e.

{ CLeaR contents of register r; INCrement contents of r; compare contents of r_i to r_j and if Equal then Jump to instruction z}

Although his model is more complicated than this simple description, the Melzak "pebble" model extended this notion of "counter" to permit multi- pebble adds and subtracts.

The Random Access Machine (RAM) Model

Melzak (1961) recognized a couple serious defects in his register/counter-machine model: (i) Without a form of indirect addressing he would not be able to "easily" show the model is Turing equivalent, (ii) The program and registers were in different "spaces", so self-modifying programs would not be easy. When Melzak added indirect addressing to his model he created a random access machine model.

(However, with Gödel numbering of the instructions Minsky (1961) offered a proof that with such numbering the general recursive functions were indeed possible; in Minsky (1967) he offers proof that μ recursion is indeed possible).

Unlike the RASP model, the RAM model does not allow the machine's actions to modify its instructions. Sometimes the model works only register-to-register with no accumulator, but most models seem to include an accumulator. van Emde Boas (1990) divides the various RAM models into a number of sub-types:

- SRAM, the "successor RAM– with only one arithmetic instruction, the successor (INCREMENT h). The others include "CLEAR h–, and an IF equality-between-register THEN jump-to xxx.

- RAM: the standard model with addition and subtraction
- MRAM: the RAM agumented with multiplication and division
- BRAM, MBRAM: Bitwise Boolean versions of the RAM and MRAM
- N****: Non-deterministic versions of any of the above with an N before the name.

The Random Access Stored Program (RASP) machine model

The RASP is a RAM with the instructions stored together with their data in the same 'space' — i.e. sequence of registers. The notion of a RASP was described at least as early as Kiphengst (1959). His model had a "mill" — an accumulator, but now the instructions were in the registers with the data—the so-called von Neumann architecture. When the RASP has alternating even and odd registers—the even holding the "operation code" (instruction) and the odd holding its "operand" (parameter), then indirect addressing is achieved by simply modifying an instruction's operand (cf Cook and Reckhow 1973).

The original RASP model of Elgot and Robinson (1964) had only three instructions in the fashion of the register-machine model, but they placed them in the register space together with their data. (Here COPY takes the place of CLEAR when one register e.g. "z" or "0" starts with and always contains 0. This trick is not unusual. The unit 1 in register "unit" or "1" is also useful.)

{ INC (r), COPY (r_i, r_j), JE (r_i, r_i, z) }

The RASP models allow indirect as well as direct-addressing; some allow "immediate" instructions too, e.g. "Load accumulator with the constant 3". The instructions may be of a highly restricted set such as the following 16 instructions of Hartmanis (1971). This model uses an accumulator A. The mnemonics are those that the authors used (their CLA is "load accumulator" with constant or from register; STO is "store accumulator"). Their syntax is the following, excepting the jumps: "n, <n>, <<n>>" for "immediate", "direct" and "indirect"). Jumps are via two "Transfer instructions" TRA—unconditional jump by directly "n" or indirectly "< n >" jamming contents of register n into the instruction counter, TRZ (conditional jump if Accumulator is zero in the same manner as TRA):

{ ADD n , ADD < n >, ADD << n >>, SUB n, SUB < n >, SUB << n >>, CLA n, CLA < n >, CLA << n >>, STO < n >, STO << n >>, TRA n, TRA < n >, TRZ n, TRA < n >, HALT }

The Pointer Machine Model

A relative late-comer is Schönhage's Storage Modification Machine (1970) or pointer machine. Another version is the Kolmogorov-Uspensii machine, and the Knuth "linking automaton" proposal. (For references see pointer machine).

Like a state-machine diagram, a node emits at least two labelled "edges" (arrows) that point to another node or nodes which in turn point to other nodes, etc. The outside world points at the center node.

Machines with Input and Output

Any of the above tape-based machines can be equipped with input and output tapes; any of the above register-based machines can be equipped with dedicated input and output registers. For example, the Schönhage pointer-machine model has two instructions called "*input* λ_0,λ_1" and "*output* β– (Schönhage 1990 p. 493)

It is difficult to study sublinear space complexity on multi-tape machines with the traditional model, because an input of size n already takes up space n.

Thus, to study small DSPACE classes, we must use a different model. In some sense, if we never "write to" the input tape, we don't want to charge ourself for this space. And if we never "read from" our output tape, we don't want to charge ourself for this space.

We solve this problem by introducing a k-string Turing machine with input and output. This is the same as an ordinary k-string Turing machine, except that the transition function δ is restricted so that the input tape can never be changed, and so that the output head can never move left.

This model allows us to define deterministic space classes smaller than linear. Turing machines with input-and-output also have the same time complexity as other Turing machines; in the words of Papaditriou 1994 Prop 2.2:

For any k-string Turing machine M operating within time bound $f(n))$ there is a $(k+2)$-string Turing machine M' with input and output, which operates within time bound $O(f(n))$.

k-string Turing machines with input and output are used in the formal definition of the complexity resource DSPACE in, for example, Papadimitriou 1994 (Def. 2.6).

Other Equivalent Machines and Methods

- Multidimensional Turing machine: For example, a model by Schönhage (1990) uses the four head-movement commands { North, South, East, West }.
- Single-tape, multi-head Turing machine: In an undecidability proof of the "problem of tag", Minsky 1961 and Shepherdson and Sturgis (1963) described machines with a single tape that could write along the tape with one head and read further along the tape with another.
- Markov's (1954) Normal Algorithm is another remarkably simple computational model equivalent to the Turing machines.
- Lambda calculus
- Queue automaton.

Chapter 6

Random Access Machine

In computer science, random access machine (RAM) is an abstract machine in the general class ofregister machines. The RAM is very similar to the counter machine but with the added capability of 'indirect addressing' of its registers. Like the counter machine the RAM has its instructions in the finite-state portion of the machine (the so-called Harvard architecture).

The RAM's equivalent of the Universal Turing machine — with its program in the registers as well as its data—is called the Random access stored program machine or RASP. It is an example of the so-called von Neumann architecture and is closest to the common notion of computer.

Together with the Turing machine and counter machine models, the RAM and RASP models are used for computational complexity analysis. Van Emde Boas (1990) calls these three plus the pointer machine "sequential machine" models, to distinguish them from "parallel random access machine" models.

Introduction to the Random Access Machine Model (RAM)

The concept of a random access machine (RAM) starts with the simplest model of all, the so-calledcounter machine model. Two additions move it away from the counter machine, however. The first enhances the machine with the convenience of indirect addressing; the second moves the model toward the more conventional accumulator-based computer with the addition of one or more auxiliary (dedicated) registers, the most common of which is called "the accumulator".

Formal Definition: Random Access Machine

A *Random Access Machine* (RAM) is an abstract computational-machine model identical to a multiple-register counter machine with the addition of indirect addressing. At the discretion of an instruction from its finite state machine's TABLE, the machine derives a "target" register's address either (i) directly from the instruction itself, or (ii) indirectly from the *contents* (e.g. number, label) of the "pointer" register specified in the instruction.

By definition: A *register* is a location with both an *address* (a unique, distinguishable designation/locator equivalent to a natural number) and a *content* — a single natural number. For precision we will use the quasi-formal symbolism from Boolos-Burgess-Jeffrey (2002) to specify a register, its contents, and an operation on a register:

- [r] means "the contents of register with address r". The label "r" here is a "variable" that can be filled with a natural number or a letter (e.g. "A") or a name.
- → means "copy/deposit into", or "replaces", but without destruction of the source.

Example: +1 → 3; means "The contents of source register with address "3", plus 1, is put into destination register with address "3" (here source and destination are the same place). If =37, that is, the contents of register 3 is the number "37", then 37+1 = 38 will be put into register 3.

Example: → 5; means "The contents of source register with address "3" is put into destination register with address "5". If =38, that is, the contents of register 3 is the number 38, then this number will be put into register 5. The contents of register 3 are not disturbed by this operation, so continues to be 38, now the same as .

Definition: A *direct* instruction is one that specifies *in the instruction itself* the address of the source or destination register whose contents will be the subject of the instruction. Definition: An *indirect instruction* is one that specifies an "pointer register", the contents of which is the address of a "target" register. The target register can be either a source or a destination (the various COPY instructions provide examples of this). A register can address itself indirectly.

For want of a standard/convention this article will specify "direct/indirect", abbreviated as "d/i", as a parameter (or parameters) in the instruction:

Example: COPY (d, A, i, N) means directly d get the source register's address (register "A") from the instruction itself but indirectly i get the destination address from pointer-register N. Suppose [N]=3, then register 3 is the destination and the instruction will do the following: [A] → 3.

Definition: The contents of *source register* is used by the instruction. The source register's address can be specified either (i) directly by the instruction, or (ii) indirectly by the pointer register specified by the instruction.

Definition: The contents of the *pointer register* is the *address* of the "target" register.

Definition: The contents of the *pointer register* points to the *target register* — the "target" may be either a source or a destination register.

Definition: The *destination register* is where the instruction deposits its result. The source register's address can be specified either (i) directly by the instruction, or (ii) indirectly by the pointer register specified by the instruction. The source and destination registers can be one

Refresher: The Counter Machine Model

Melzak (1961) provides an easy visualization of a counter machine: its "registers" are holes in the ground, and these holes hold pebbles. Per an instruction, into and out of these holes "the computer" (person or machine) adds (INCrements) or removes (DECrements) a single pebble. As needed, additional pebbles come from, and excess pebbles go back into, an infinite supply; if the hole is too small to accommodate the pebbles the "computer" digs the hole bigger.

Minsky (1961) and Hopcroft-Ullman 1979 (p. 171) offer the visualization of a multi-tape Turing machine with as many left-ended tapes as "registers".

Each tape's length is unbounded to the right, and every square is blank except for the left end, which is marked. The distance of a tape's "head" from its left end, measured in numbers of tape-squares, represents the natural number in "the register". To DECrement the count of squares the tape head moves left; INCrement it moves right. There is no need to print or erase marks on the tape; the only conditional instructions are to check to see if the head is at the left end, by testing a left-end mark with a "Jump-if-marked instruction".

The following instruction "mnemonics" e.g. "CLR (r)" are arbitrary; no standard exists.

The register machine has, for a memory external to its finite-state machine—an unbounded (cf: footnote|countable and unbounded) collection of discrete and uniquely labelled locations with*unbounded* capacity, called "registers". These registers hold only natural numbers (zero and the positive integers). Per a list of sequential instructions in the finite state machine's TABLE, a few (e.g. 2) types of primitive operations operate on the contents of these "registers". Finally, a *conditional-expression* in the form of an *IF-THEN-ELSE* is available to test the contents of one or two registers and "branch/jump" the finite state machine out of the default instruction-sequence.

Base model 1: The model closest to Minsky's (1961) visualization and to Lambek (1961):

{ INCrement contents of register r, DECrement contents of register r, *IF* contents of register r is Zero *THEN* Jump to instruction I_z *ELSE* continue to next instruction }:

Instruction	Mnemonic	Action on register(s) "r"	Action on finite state machine's Instruction Register, IR
INCrement	INC (r)	[r] +1 ? r	[IR] +1 ? IR
DECrement	DEC (r)	[r] -1 ? r	[IR] +1 ? IR
Jump if Zero	JZ (r, z)	none	IF [r1] = 0 THEN z ? IR ELSE [IR] +1 ? IR
Halt	H	none	[IR] ? IR

Base model 2: The "successor" model (named after the successor function of the Peano axioms):

- { INCrement the contents of register r, CLeaR the contents of register r, *IF* contents of register r_j Equals the contents of register r_k *THEN* Jump to instruction I_z *ELSE* goto to next instruction }

Instruction	Mnemonic	Action on register(s) "r"	Action on finite state machine's Instruction Register, IR
CLeaR	CLR (r)	0 → r	[IR] +1 → IR
INCrement	INC (r)	[r] +1 → r	[IR] +1 → IR
Jump if Equal	JE (r1, r2, z)	none	IF [r1] = [r2] THEN z → IR ELSE [IR] +1 → IR
Halt	H	none	[IR] → IR

Base model 3: Used by Elgot-Robinson (1964) in their investigation of bounded and unbounded RASPs—the "successor" model with COPY in the place of CLEAR:

- { INCrement the contents of register r, COPY the contents of register r_j to register r_k, *IF* contents of register r_j Equals the contents of register r_k *then* Jump to instruction I_z *ELSE* goto to next instruction }

Creating "Convenience Instructions" From the Base Sets

The three base sets 1, 2, or 3 above are equivalent in the sense that one can create the instructions of one set using the instructions of another set (an interesting exercise: a hint from Minsky (1967) — declare a reserved register e.g. call it "0" (or Z for "zero" or E for "erase") to contain the number 0). The choice of model will depend on which an author finds easiest to use in a demonstration, or a proof, etc.

Moreover, from base sets 1, 2, or 3 we can create *any* of the primitive recursive functions (cf Minsky (1967), Boolos-Burgess-Jeffrey (2002)). (How to cast the net wider to capture the *total* and *partial* mu recursive functions will be discussed in context of indirect addressing). However, building the primitive recursive functions is difficult because the instruction sets are so ... primitive (tiny). One solution is to expand a particular set with "convenience instructions" from another set:

These will not be subroutines in the conventional sense but rather blocks of instructions created from the base set and given a mnemonic. In a formal sense, to use these blocks we need to either (i) "expand" them into their base-instruction equivalents — they will require the use of temporary or "auxiliary" registers so the model must take this into account, or (ii) design our machines/models with the instructions 'built in'.

Example: Base set 1. To create CLR (r) use the block of instructions to count down register r to zero. Observe the use of the hint mentioned above:

- CLR (r) $=_{\text{equiv}}$
- *loop*: JZ (r, *exit*)
- DEC (r)
- JZ (0, *loop*)
- *exit*: etc.

Again, all of this is for convenience only; none of this increases the model's intrinsic power.

For example: the most expanded set would include each unique instruction from the three sets, plus unconditional jump J (z) i.e.:

- { CLR (r), DEC (r), INC (r), CPY (r_s, r_d), JZ (r, z), JE (r_j, r_k, z), J(z) }

Most authors pick one or the other of the conditional jumps, e.g. Shepherdson-Sturgis (1963) use the above set minus JE (to be perfectly accurate they use JNZ—Jump if *Not* Zero in place of JZ; yet another possible convenience instruction).

The "indirect" Operation

Example of Indirect Addressing

In our daily lives the notion of an "indirect operation" is not unusual.

Example: A treasure hunt —

At location "Tom_&_Becky's_cave_in_pirate_chest" will be where we can find a map directing us to "the treasure":

(1) We go to location "Tom_&_Becky's_cave..." and dig around until we find a wooden box

(2) Inside the box is a map to the location of the treasure: "under_Thatcher's_front_porch"

(3) We go to location "under_Thatcher's_front_porch", jackhammer away the concrete, and discover "the treasure": a sack of rusty door-knobs.

Indirection specifies a location identified as the pirate chest in "Tom_&_Becky's_cave..." that acts as a *pointer* to any other location (including itself): its contents (the treasure map) provides the "address" of the *target* location "under_Thatcher's_front_porch" where the real action is occurring.

Why the Need for an Indirect Operation: Two Major Problems with the Counter Machine Model

In the following one must remember that these models are abstract models with two fundamental differences from anything physically real: unbounded numbers of registers each with unbounded capacities. The problem appears most dramatically when one tries to use a counter machine model to build a RASP (Random access stored program machine) that is Turing equivalent and thus compute any partial mu recursive function:

- Melzak (1961) added indirection to his "hole-and-pebble" model so that his model could modify itself with a "computed goto" and provides two examples of its use ("Decimal representation in the scale of ≤ and "Sorting by magnitude", whether these are used in his proof that the model is Turing equivalent is unclear since "the program iteself is left to the reader as an exercise" (p. 292)). Minsky (1961, 1967) was able to demonstrate that, with suitable (but difficult-to-use) Gödel numberencoding, the register model did not need indirection to be Turing equivalent; but it did need at least one unbounded register. As noted below, Minsky (1967) hints at the problem for a RASP but doesn't offer a solution. Elgot and Robinson (1964) proved that their RASP model P_0 — it has no indirection capability— cannot compute all "recursive sequential functions" (ones that have parameters of arbitrary length) if it does not have the capability of modifying its own instructions, but it can via Gödel numbers if it does (p. 395-397; in particular figure 2 and footnote p. 395). On the other hand their RASP model P'_0 equipped with an "index register" (indirect addressing) can compute all the "partial recursive sequential functions" (the mu recursive functions) (p. 397-398).

Cook and Reckhow (1973) Say it Most Succinctly: The indirect instructions are necessary in order for a fixed progam to access an unbounded number of registers as the inputs vary." (p. 73)

- *Unbounded Capacities of Registers Versus Bounded Capacities of State-machine Instructions:* The so-called *finite* state part of the machine is supposed to be—by the normal definition of algorithm — *very* finite both in the number of "states" (instructions) and the instructions' sizes (their capacity to hold symbols/signs). So how does a state machine move an arbitrarily large constant directly into a register, e.g. MOVE (k, r) (Move constant k to register r)? If huge constants are necessary they must either start out in the registers themselves or be created by the state machine using a finite number of instructions e.g. multiply and add subroutines using INC and DEC (but not a quasi-infinite number of these!).

Sometimes the constant k will be created by use of CLR (r) followed by INC (r) repeated k times — e.g. to put the constant k=3 into register r, i.e. 3 → r, so at the end of the instruction [r]=3: CLR (r), INC (r), INC (r), INC (r). This trick is mentioned by Kleene (1952) p. 223. The problem arises when the number to be created exhausts the number of instructions available to the finite state machine; there is always a bigger constant than the number of instructions available to the finite state machine.

- *Unbounded numbers of registers versus bounded state-machine instructions:* This is more severe than the first problem. In particular, this problem arises when we attempt to build a so-called Random Access Stored Program machine (RASP), a "Universal Machine" (see more at Universal Turing machine) that uses its finite state machine to interpret a "program of instructions" located in its registers—i.e. we are building what is nowadays called a computer with the von Neumann architecture.

Observe that the counter machine's finite state machine must call out a register *explicitly* (directly) by its name/number: INC (65,356) calls out register number "65,365" *explicitly*. If the number of registers exceeds the capability of the *finite* state machine to address them, then registers outside the bounds will be unreachable. For example, if the finite state machine can only reach 65,536 = 2^{16} registers then how can it reach the 65,537th?

So how *do* we address a register beyond the bounds of the finite state machine? One approach would be to modify the *program*-instructions (the ones stored in the registers) so that they contain more than one command. But this too can be exhausted unless an instruction is of (potentially) unbounded size. So why not use just one "über-instruction"—one really really big number—that contains *all* the program instructions encoded into it! This is how Minsky solves the problem, but the Gödel numberinghe uses represents a great inconvenience to the model, and the result is nothing at all like our intuitive notion of a "stored program computer".

Elgot and Robinson (1964) come to a similar conclusion with respect to a RASP that is "finitely determined". Indeed it can access an unbounded number of registers (e.g. to fetch instructions from them) but only if the RASP allows "self modification" of its *program* instructions, and has encoded its "data" in a Gödel number (Fig. 2 p. 396).

In the context of a more computer-like model using his RPT (repeat) instruction Minsky (1967) tantalizes us with a solution to the problem (cf p. 214, p. 259) but offers no firm resolution. He asserts:

> *"In general a RPT operation could not be an instruction in the finite-state part of the machine ... this might exhaust any particular amount of storage allowed in the finite part of the computer [sic, his name for his RAM*

models]. RPT operations require infinite registers of their own." (p. 214).

He offers us a *bounded* RPT that together with CLR (r) and INC (r) can compute any primitive recursive function, and he offers the unbounded RPT quoted above that as playing the role of μ operator; it together with CLR (r) and INC (r) can compute the mu recursive functions. But he does not discuss "indirection" or the RAM model per se. From the references in Hartmanis (1971) it appears that Cook (in his lecture notes while at UC Berkeley, 1970) has firmed up the notion of indirect addressing. This becomes clearer in the paper of Cook and Reckhow (1973) — Cook is Reckhow's Master's thesis advisor. Hartmanis' model—quite similar to Melzak's (1961) model—uses two and three-register adds and subtracts and two parameter copies; Cook and Reckhow's model reduce the number of parameters (registers called out in the program instructions) to one call-out by use of an accumulator "AC".

The solution in a nutshell: Design our machine/model with unbounded indirection — provide an*unbounded* "address" register that can potentially name (call out) any register no matter how many there are. For this to work, in general, the *unbounded* register requires an ability to be cleared and then incremented (and, possibly, decremented) by a potentially infinite loop. In this sense the solution represents the unbounded μ operator that can, if necessary, hunt ad infinitim along the unbounded string of registers until it finds what it is looking for. The pointer register is exactly like any other register with one exception: under the circumstances called "indirect addressing" it provides *its*contents, rather than the address-operand in the state machine's TABLE, to be the address of the target register (including possibly itself!)

Bounded Indirection and the Primitive Recursive Functions

If we eschew the Minsky approach of one monster number in one register, and specify that our machine model will be "like a computer" we have to confront this problem of indirection if we are to compute the recursive functions (also called the μ-recursive functions) — both total and partial varieties.

Our simpler counter machine model can do a "bounded" form of indirection—and thereby compute the sub-class of primitive recursive functions — by using a primitive recursive "operator" called "definition by cases" (defined in Kleene (1952) p. 229 and Boolos-Burgess-Jeffrey p. 74). Such a "bounded indirection" is a laborious, tedious affair. "Definition by cases" requires the machine to determine/distinguish the contents of the pointer register by attempting, time after time until success, to match this contents against a number/name that the case operator *explicitly* declares. Thus the definition by cases starts from e.g. the lower bound address and continues ad nauseam toward the upper bound address attempting to make a match:

> *Is the number in register N equal to 0? If not then is it equal to 1? 2? 3? ... 65364? If not then we're at the last number 65365 and this had better be the one, else we have a problem!*
>
> *"Bounded" indirection will not allow us to compute the partial recursive functions—for those we needunbounded indirection aka the μ operator.*
>
> *Suppose we had been able to continue on to number 65367, and in fact that register had what we were looking for. Then we could have completed our calculation successfully! But suppose 65367 didn't have what we needed. How far should we continue to go?*

To be Turing equivalent the counter machine needs to either use the unfortunate single-register MinskyGödel number method, or be augmented with an ability to explore the ends of its register string, ad infinitum if necessary. (A failure to find something "out there" defines what it means for an algorithm to fail to terminate; cf Kleene (1952) pp. 316ff *Chapter XII Partial Recursive Functions*, in particular p. 323-325.) See more on this in the example below.

Unbounded Indirection and the Partial Recursive Functions

For *unbounded* indirection we require a "hardware" change in our machine model. Once we make this change the model is no longer a counter machine, but rather a random access machine.

Now when e.g. INC is specified, the finite state machine's instruction will have to specify *where* the address of the register of interest will come from. This *where* can be either (i) the state machine's instruction that provides an *explicit label*, or (ii) the *pointer-register* whose *contents* is the address of interest. Whenever an instruction specifies a register address it now will *also* need to specify an additional parameter "i/d-— "indirect/direct". In a sense this new "i/d- parameter is a "switch" that flips one way to get the direct address as specified in the instruction or the other way to get the indirect address from the pointer register (which pointer register—in some models every register can be a pointer register—is specified by the instruction). This "mutually exclusive but exhaustive choice" is yet another example of "definition by cases", and the arithmetic equivalent shown in the example below is derived from the definition in Kleene (1952) p. 229.

Example: CPY ($indirect_{source}$, r_{source}, $direct_{destination}$, $r_{destination}$)

Assign a code to specify direct addressing as d="0– and indirect addressing as i="1–. Then our machine can determine the source address as follows:

$i*[r_s] + (1-i)*r_s$

For example, suppose the contents of register 3 are "5" (i.e. =5) and the contents of register 4 are "2" (i.e. =2):

Example: CPY (1, 3, 0, 4) = CPY (indirect, reg 3, direct, reg 4)

1* + 0*3 = = source-register address 5

0* + 1*4 = 4 = destination-register address 4

Example: CPY (0, 3, 0, 4)

0* + 1*3 = 3 = source-register address 3

0* + 1*4 = 4 = destination-register address 4

Example: CPY (0, 3, 1, 4)

0* + 1*3 = 3 = source-register address 3

1* + 0*4 = = destination-register address 2

The Indirect COPY Instruction

Probably the most useful of the added instructions is COPY. Indeed Elgot-Robinson (1964) provide their models P_0 and P'_0 with the COPY instructions, and Cook-Reckhow (1973) provide their accumulator-based model with only two indirect instructions—COPY to accumulator indirectly, COPY from accumulator indirectly.

A plethora of instructions: Because any instruction acting on a single register can be augmented with its indirect "dual" (including conditional and unconditional jumps, cf the Elgot-Robinson model), the inclusion of indirect instructions will double the number of single parameter/register instructions (e.g. INC (d, r), INC (i, r)). Worse, every two parameter/register instruction will have 4 possible varieties, e.g.:

CPY (d, r_s, d, r_d) = COPY directly from source-register directly to destination-register

CPY (i, r_{sp}, d, r_d) = COPY to destination-register indirectly using the source address to be found in the source-pointer register r_{sp}.

CPY (d, r_s, i, r_{dp}) = COPY contents of source-register indirectly into register using destination address to be found in the destination-pointer register r_{dp}.

CPY (i, r_{sp}, i, r_{dp}) = COPY indirectly the contents of the source register with address to be found in source-pointer register r_{sp}, into the destination register with address to be found in the destination-pointer register r_{dp})

In a similar manner every three-register instruction that involves two source registers r_{s1} r_{s2} and a destination register r_d will result in 8 varieties, for example the addition:

$[r_{s1}] + [r_{s2}] \rightarrow r_d$

will yield:

- ADD (d, r_{s1} , d, r_{s2} , d, r_d)
- ADD (i, r_{sp1} , d, r_{s2} , d, r_d)
- ADD (d, r_{s1} , i, r_{sp2} , d, r_d)
- ADD (i, r_{sp1} , i, r_{sp2} , d, r_d)
- ADD (d, r_{s1} , d, r_{s2} , i, r_{dp})
- ADD (i, r_{sp1} , d, r_{s2} , i, r_{dp})
- ADD (d, r_{s1} , i, r_{sp2} , i, r_{dp})
- ADD (i, r_{s1} , i, r_{s2} , i, r_{dp}).

If we designate one register to be the "accumulator" (see below) and place strong restrictions on the various instructions allowed then we can greatly reduce the plethora of direct and indirect operations. However, one must be sure that the resulting reduced instruction-set is sufficient, and we must be aware that the reduction will come at the expense of more instructions per "significant" operation.

The Notion of "Accumulator A"

Historical convention dedicates a register to the accumulator, an "arithmetic organ" that literally accumulates its number during a sequence of arithmetic operations:

"The first part of our arithmetic organ ... should be a parallel storage organ which can receive a number and add it to the one already in it, which is also able to clear its contents and which can store what it contains. We will call such an organ an *Accumulator*. It is quite conventional in principle in past and present computing machines of the most varied types, e.g. desk multipliers, standard IBM counters, more modern relay machines, the ENIAC" (boldface in original: Goldstine and von Neumann, 1946; p. 98 in Bell and Newell 1971).

However, the accumulator comes at the expense of more instructions per arithmetic "operation", in particular with respect to what are called 'read-modify-write' instructions such as "Increment indirectly the contents of the register pointed to by register r2 ". "A" designates the "accumulator" register A:

Label	Instruction		A	r2	r378,426	Description
			. . .	378,426	17	
INCi (r2):	CPY (i, r2, d, A)		**17**	378,426	17	Contents of r2 points to r378,426 with contents "17": copy this to A
	INC (A)		**18**	378,426	17	Incement contents of A
	CPY (d, A, i, r2)		18	378,426	**18**	Contents of r2 points to r378,426: copy contents of A into r378,426

If we stick with a specific name for the accumulator, e.g. "A", we can imply the accumulator in the instructions, for example,

INC (A) = INCA

However, when we write the CPY instructions without the accumulator called out the instructions are ambiguous or they must have empty parameters:

CPY (d, r2, d, A) = CPY (d, r2, ,)

CPY (d, A, d, r2) = CPY (, , d, r2)

Historically what has happened is these two CPY instructions have received distinctive names; however, no convention exists. Tradition (e.g. Knuth's (1973) imaginary MIX computer) uses two names called LOAD and STORE. Here we are adding the "i/≤ parameter:

LDA (d/i, r_s) $=_{def}$ CPY (d/i, r_s, d, A)

STA (d/i, r_d) $=_{def}$ CPY (d, A, d/i, r_d)

The typical accumulator-based model will have all its two-variable arithmetic and constant operations (e.g. ADD (A, r), SUB (A, r)) use (i) the accumulator's contents, together with (ii) a specified register's contents. The one-variable operations (e.g. INC (A), DEC (A) and CLR

(A)) require only the accumulator. Both instruction-types deposit the result (e.g. sum, difference, product, quotient or remainder) in the accumulator.

Example: INCA = [A] +1 → A

Example: ADDA (r_s) = [A] + [r_s] → A

Example: MULA (r_s) = [A] * [r_s] → A

If we so choose, we can abbreviate the mnemonics because at least one source-register and the destination register is always the accumulator A. Thus we have :

{ LDA (i/d, r_s), STA (i/d, r_d), CLRA, INCA, DECA, ADDA (r_s), SUBA (r_s), MULA (r_s), DIVA (r_s), etc.)

The Notion of Indirect Sddress Register "N"

If our model has an *unbounded accumulator* can we *bound* all the other registers? Not until we provide for at least one unbounded register from which we derive our indirect addresses.

The minimimalist approach is to use itself (Schönhage does this).

Another approach (Schönhage does this too) is to declare a specific register the "indirect address register" and confine indirection relative to this register (Schonhage's RAM0 model uses both A and N registers for indirect as well as direct instructions). Again our new register has no conventional name—perhaps "N" from "iNdex", or "iNdirect" or "address Number".

For maximum flexibility, as we have done for the accumulator A—we will consider N just another register subject to increment, decrement, clear, test, direct copy, etc. Again we can shrink the instruction to a single-parameter that provides for direction and indirection, for example.

LDAN (i/d) = CPY (i/d, N, d, A); LoaD Accumulator via iNdirection register

STAN (i/d) = CPY (d, A, i/d, N). STore Accumlator via iNdirection register

Why is this such an interesting approach? At least two reasons:

(1) An instruction set with no parameters:

Schönhage does this to produce his RAM0 instruction set. See section below.

(2) Reduce a RAM to a Post-Turing machine:

Posing as minimalists, we reduce all the registers excepting the accumulator A and indirection register N e.g. r = { r0, r1, r2, ... } to an unbounded string of (very-) bounded-capacity pigeon-holes. These will do nothing but hold (very-) bounded numbers e.g. a lone bit with value { 0, 1 }. Likewise we shrink the accumulator to a single bit. We restrict any arithmetic to the registers { A, N }, use indirect operations to pull the contents of registers into the accumulator and write 0 or 1 from the accumulator to a register:

{ LDA (i, N), STA (i, N), CLR (A/N), INC (A/N), DEC(N), JZ (A/N, I_z), JZ (I_z), H }

We push further and eliminate A altogether by the use of two "constant" registers called "ERASE" and "PRINT": [ERASE]=0, [PRINT]=1.

{ CPY (d, ERASE, i, N), CPY (d, PRINT, i, N), CLR (N), INC (N), DEC (N), JZ (i, N, I_z), JZ (I_z), H }

Rename the COPY instructions and call INC (N) = RIGHT, DEC (N) = LEFT and we have the same instructions as the Post-Turing machine, plus an extra CLRN :

{ERASE, PRINT, CLRN, RIGHT, LEFT, JZ (i, N, I_z), JZ (I_z), H }

Turing Equivalence of the RAM with Indirection

In the section above we informally showed that a RAM with an unbounded indirection capability produces a Post-Turing machine. The Post-Turing machine is Turing equivalent, so we have shown that the RAM with indirection is Turing equivalent.

We give here a slightly more formal demonstration. Begin by designing our model with three reserved registers "E", "P", and "N", plus an unbounded set of registers 1, 2, ..., n to the right. The registers 1, 2, ..., n will be considered "the squares of the tape". Register "N" points to "the scanned square" that "the head" is currently observing. The "head" can be thought of as being in the conditional jump—observe that it uses indirect addressing (cf Elgot-Robinson p. 398). As we decrement or increment "N" the (apparent) head will "move left" or "right" along the squares. We will move the contents of "E"=0 or "P"=1 to the "scanned square" as pointed to by N, using the indirect CPY.

The fact that our tape is left-ended presents us with a minor problem: Whenever LEFT occurs our instructions will have to test to determine whether or not the contents of "N" is zero; if so we should leave its count at "0" (this is our choice as designers—for example we might have the machine/model "trigger an event" of our choosing).

Instruction set 1 (augmented): { INC (N), DEC (N), CLR (N), CPY (d, r_s,i, N), JZ (i, r, z), HALT }

The following table both defines the Post-Turing instructions in terms of their RAM equivalent instructions and gives an example of their functioning. The (apparent)location of the head along the tape of registers r0-r5 . . . is shown shaded:

Example: Bounded indirection yields a machine that is not Turing equivalent

Throughout this demonstration we have to keep in mind that the instructions in the finite state machine's TABLE is *bounded,* i.e. *finite*:

"Besides a merely being a *finite set of rules* which gives a seqeunce of operations for solving a specific type of problem, an algorithm has five important features [Finiteness, Definiteness, Input, Output, Effectiveness]" (italics added, Knuth p. 4-7).

The difficulty arises because the registers have explicit "names" (numbers) and our machine must call each out by name in order to "access" it.

We will build the indirect CPY (i, q, d, φ) with the CASE operator.

The address of the target register will be specified by the contents of register "q"; once the CASE operator has determined what this number is, CPY will directly deposit the contents of the register with that number into register "φ". We will need an additional register that we will call "y" — it serves as an up-counter.

So the following is actually a constructive demonstration or proof that we can indeed simulate the indirect CPY (i, q, d, φ) without a "hardware" design change to our counter machine/model. However, note that because this indirect CPY is "bounded" by the size/extent of the finite state machine, a RASP using this indirect CPY can only calculate the primitive recursive functions, not the full suite of mu recursive functions.

The CASE "operator" is described in Kleene (1952) (p. 229) and in Boolos-Burgess-Jeffrey (2002) (p. 74); the latter authors emphasize its utility. The following definition is per Kleene but modified to reflect the familiar "IF-THEN-ELSE" construction.

The CASE operator "returns" a natural number into φ depending on which "case" is satisfied, starting with "case_0" and going successively through "case_last"; if no case is satisfied then the number called "default" (aka "woops") is returned into φ (here x designates some selection of parameters, e.g. register q and the string r0, ... rlast)):

Definition by cases φ (x, y):

- case_0: IF Q_0(x, y) is true THEN φ_0(x, y) ELSE
- case_1: IF Q_1(x, y) is true THEN φ_1(x, y) ELSE
- cases_2 through case_next_to_last: etc. . . . . ELSE
- case_last: IF Q_{last}(x, y) is true THEN φ_{last}(x, y) ELSE
- default: do $\varphi_{default}$(x, y).

Kleene require that the "predicates" Q_n that doing the testing are all mutually exclusive — "predicates" are functions that produce only { true, false } for output; Boolos-Burgess-Jeffrey add the requirement that the cases are "exhaustive".

We begin with a number in register q that represents the address of the target register. But what is this number? The "predicates" will test it to find out, one trial after another: JE (q, y, z) followed by INC (y). Once the number is identified explicitly, the CASE operator directly/explicitly copies the contents of this register to φ:

Definition by cases CPY (i, q, d, φ) $=_{def}$ φ (q, r0, ..., rlast, y) =

- case_0: IF CLR (y), [q] - [y]=0 THEN CPY (r0, φ), J (exit) ELSE
- case_1: IF INC (y), [q] = [y]=1 THEN CPY (r1, φ), J (exit) ELSE
- case_2 through case n: IF . . . THEN . . . ELSE
- case_n: IF INC (y), [q] = [y]=n THEN CPY (rn, φ), J (exit) ELSE
- case_n+1 to case_last: IF . . . THEN . . . ELSE
- case_last: IF INC (y), [q] = [y]="last" THEN CPY (rlast, φ), J (exit) ELSE
- default: woops.

Case_0 (the base step of the recursion on y) looks like this:

- *case_0*:
- CLR (y) ; set register y = 0
- JE (q, y, *_φ0*)

- J (*case_1*)
- *_φ0:* CPY (r0, φ)
- J (*exit*)
- *case_1:* etc.

Case_n (the induction step) looks like this; remember, each instance of "n", "n+1", ..., "last" must be an explicit natural number:

- *case_n*:
- INC (y)
- JE (q, y, *_φn*)
- J (*case_n+1*)
- *_φn:* CPY (rn, φ)
- J (*exit*)
- *case__n+1:* etc.

Case_last stops the induction and bounds the CASE operator (and thereby bounds the "indirect copy" operator):

- *case_last*:
- INC (y)
- JE (q, y, *_φlast*)
- J (*woops*)
- *_φlast*: CPY (rlast, φ)
- J (*exit*)
- *woops:* how do we handle an out-of-bounds attempt?
- *exit:* etc.

If the CASE could continue ad infinitum it would be the mu operator. But it can't—its finite state machine's "state register" has reached its maximum count (e.g. 65365 = $11111111,11111111_2$) or its table has run out of instructions; it is a *finite* machine, after all.

Examples of Models

Register-to-register ("read-modify-write") model of Cook and Reckhow (1973)

The commonly encountered Cook and Rechkow model is a bit like the ternary-register Malzek model (written with Knuth mnemonics—the original instructions had no mnemonics excepting TRA, Read, Print).

- LOAD (C, r_d) ; C → r_d, C is any integer

Example: LOAD (0, 5) will clear register 5.

- ADD (r_{s1}, r_{s2}, r_d) ; [r_{s1}] + [r_{s2}] → r_d, the registers can be the same or different;

Example: ADD (A, A, A) will double the contents of register A.

- SUB (r_{s1}, r_{s2}, r_d) ; [r_{s1}] - [r_{s2}] → r_d, the registers can be the same or different:

Example: SUB (3, 3, 3) will clear register 3.

- COPY (i, r_p, d, r_d) ; [[r_p]] → r_d, Indirectly copy the contents of the source-register pointed to by pointer-register r_p into the destination register.
- COPY (d, r_s, i, r_p) ; [r_s] → [r_p]. Copy the contents of source register r_s into the destination-register pointed to by the pointer-register r_p.
- JNZ (r, I_z) ; Conditional jump if [r] is positive; i.e. IF [r] > 0 THEN jump to instruction z else continue in sequence (Cook and Reckhow call this: "TRAnsfer control to line m if Xj > 0")
- READ (r_d) ; copy "the input" into destination register r_d
- PRINT (r_s) ; copy the contents of source register r_s to "the output."

Schönhage's RAM0 and RAM1 (1980)

Schönhage (1980) describes a very primitive, atomized model chosen for his proof of the equivalence of his SMM pointer machine model:

> *"In order to avoid any explicit addressing the RAM0 has the accumulator with contents* z *and an additional address register with current contents* n *(initially 0)" (p. 494)*

RAM1 model: Schönhage demonstrates how his construction can be used to form the more common, usable form of "successor"-like RAM (using this article's mnemonics):

- LDA k ; k —> A , k is a constant, an explicit number such as "47"
- LDA (d, r) ; [r] → A ; directly load A
- LDA (i, r) ; [[r]] → A ; indirectly load A
- STA (d, r) ; [A] → r ; directly store A

- STA (i, r) ; [A] → [r] ; indirectly store A
- JEA (r, z) ; IF [A] = [r] then I_z else continue
- INCA ; [A] + 1 —> A.

RAM0 model: Schönhage's RAM0 machine has 6 instructions indicated by a single letter (the 6th "C xxx" seems to involve 'skip over next parameter'. Schönhage designated the accumulator with "z", "N" with "n", etc. Rather than Schönage's mnemonics we will use the mnemonics developed above.

- (Z), CLRA: 0 → A
- (A), INCA: [A] +1 → A
- (N), CPYAN: [A] → N
- (A), LDAA: [[A]] → A ; contents of A points to register address; put register's contents into A
- (S), STAN: [A] → [N] ; contents of N points to register address; put contents of A into register pointed to by N
- (C), JAZ (z): [A] = 0 then go to I_z ; ambiguous in his treatment

Indirection comes (i) from CPYAN (copy/transfer contents A to N) working with store_A_via_N STAN, and from (ii) the peculiar indirection instruction LDAA ([[A]] → [A]).

Footnotes

Finite vs Unbounded

The definitional fact that any sort of counter machine without an unbounded register-"address" register must specify a register "r" by name indicates that the model requires "r" to be *finite*, although it is "unbounded" in the sense that the model implies no upper limit to the number of registers necessary to do its job(s). For example we do not require r < 83,617,563,821,029,283,746 nor r < 2^1,000,001, etc. Thus our model can "expand" the number of registers, if necessary to perform a certain computation. However this *does* mean that whatever number the model expands to must be *finite* — it must be indexable with a natural number: *ω is not an option.* We can escape this restriction by providing an unbounded register to provide the address of the register that specifies an indirect address.

Non-deterministic Turing Machine

In theoretical computer science, a Turing machine is a theoretical machine that is used in thought experiments to examine the abilities and limitations of computers. In essence, a Turing machine is imagined to be a simple computer that reads and writes symbols one at a time on an endless tape by strictly following a set of rules.

It determines what action it should perform next according to its internal "state" and what symbol it currently sees. An example of one of a Turing Machine's rules might thus be: "If you are in state 2 and you see an 'A', change it to a 'B' and move left."

In a deterministic Turing machine, the set of rules prescribes at most one action to be performed for any given situation. A non-deterministic Turing machine (NTM), by contrast, may have a set of rules that prescribes more than one action for a given situation.

For example, a non-deterministic Turing machine may have both "If you are in state 2 and you see an 'A', change it to a 'B' and move left" and "If you are in state 2 and you see an 'A', change it to a 'C' and move right" in its rule set.

An ordinary (deterministic) Turing machine (DTM) has a transition function that, for a given state and symbol under the tape head, specifies three things: the symbol to be written to the tape, the direction (left or right) in which the head should move, and the subsequent state of the finite control.

For example, an X on the tape in state 3 might make the DTM write a Y on the tape, move the head one position to the right, and switch to state 5.

A non-deterministic Turing machine (NTM) differs in that the state and tape symbol no longer *uniquely*specify these things; rather, many different actions may apply for the same combination of state and symbol. For example, an X on the tape in state 3 might now allow the NTM to write a Y, move right, and switch to state 5 *or* to write an X, move left, and stay in state 3.

Definition

A nondeterministic Turing machine can be formally defined as a 6-tuple $M = (Q,\Sigma,\iota,\sqcup,A,\delta)$, where

- Q is a finite set of states
- Σ is a finite set of symbols (the tape alphabet)
- $\iota \in Q$ is the initial state
- $\sqcup \in \Sigma$ is the blank symbol
- $A \subseteq Q$ is the set of accepting states

- $\delta \subseteq (Q \setminus A \times \Sigma) \times (Q \times \Sigma \times \{L,R\})$ is a relation on states and symbols called the*transition relation.*

The difference with a standard (deterministic) Turing machine is that for those, the transition relation is a function (the transition function).

Configurations and the *yields* relation on configurations, which describes the possible actions of the Turing machine given any possible contents of the tape, are as for standard Turing machines, except that the yields relation is no longer single-valued.

The notion of string acceptance is unchanged: a non-deterministic Turing machine accepts a string if, when the machine is started on the configuration in which the tape head is on the first character of the string (if any), and the tape is all blank otherwise, at least one of the machine's possible computations from that configuration puts the machine into a state in A. (If the machine is deterministic, the possible computations are the prefixes of a single, possibly infinite, path.)

Resolution of Multiple Rules

How does the NTM "know" which of these actions it should take? There are two ways of looking at it. One is to say that the machine is the "luckiest possible guesser"; it always picks the transition which eventually leads to an accepting state, if there is such a transition. The other is to imagine that the machine "branches" into many copies, each of which follows one of the possible transitions. Whereas a DTM has a single "computation path" that it follows, an NTM has a "computation tree". If any branch of the tree halts with an "accept" condition, we say that the NTM accepts the input.

Variations

Equivalence with DTMs

In particular, nondeterministic Turing machines are equivalent with deterministic Turing machines. This equivalency refers to what can be computed, as opposed to how quickly.

NTMs effectively include DTMs as special cases, so it is immediately clear that DTMs are not more powerful. It might seem that NTMs are more powerful than DTMs, since they can allow trees of possible computations arising from the same initial configuration, accepting a string if any one branch in the tree accepts it.

However, it is possible to simulate NTMs with DTMs: One approach is to use a DTM of which the configurations represent multiple configurations of the NTM, and the DTM's operation consists of visiting each of them in turn, executing a single step at each visit, and spawning new configurations whenever the transition relation defines multiple continuations.

Another construction simulates NTMs with 3-tape DTMs, of which the first tape always holds the original input string, the second is used to simulate a particular computation of the NTM, and the third encodes a path in the NTM's computation tree. The 3-tape DTMs are easily simulated with a normal single-tape DTM. In this construction, the resulting DTM effectively performs a breadth-first search of the NTM's computation tree, visiting all possible computations of the NTM in order of increasing length until it finds an accepting one. Therefore, the length of an accepting computation of the DTM is, in general, exponential in the length of the shortest accepting computation of the NTM. This is considered to be a general property of simulations of NTMs by DTMs; the most famous unresolved question in computer science, the P = NP problem, is related to this issue.

Bounded Non-determinism

An NTM has the property of bounded non-determinism, *i.e.*, if an NTM always halts on a given input tape *T* then it halts in a bounded number of steps, and therefore can only have a bounded number of possible configurations.

Comparison with Quantum Computers

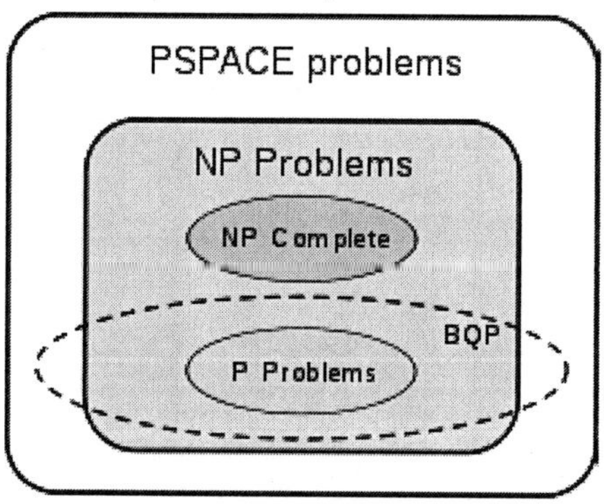

Figure: The suspected shape of the range of problems solvable by quantum computers in polynomial time (BQP). Note that the figure suggests $\mathbf{P \neq NP}$ and $\mathbf{NP \neq PSPACE}$. If this is not true then the figure should look different.

It is a common misconception that quantum computers are NTMs. It is believed but has not been proven that the power of quantum computers is incomparable to that of NTMs. That is, problems likely exist that an NTM could efficiently solve but that a quantum computer cannot. A likely example of problems solvable by NTMs but not by quantum computers in polynomial time are NP-completeproblems.

Complexity Measures

For a precise definition of what it means to solve a problem using a given amount of time and space, a computational model such as the deterministic Turing machine is used. The *time required* by a deterministic Turing machine M on input x is the total number of state transitions, or steps, the machine makes before it halts and outputs the answer ("yes" or "no"). A Turing machine M is said to operate within time $f(n)$, if the time required by M on each input of length n is at most $f(n)$. A decision problem A can be solved in time $f(n)$ if there exists a Turing machine operating in time $f(n)$ that solves the problem. Since complexity theory is interested in classifying problems based on their difficulty, one defines sets of problems based on some criteria. For instance, the set of problems solvable within time $f(n)$ on a deterministic Turing machine is then denoted by DTIME($f(n)$).

Analogous definitions can be made for space requirements. Although time and space are the most well-known complexity resources, any complexity measure can be viewed as a computational resource. Complexity measures are very generally defined by the Blum complexity axioms. Other complexity measures used in complexity theory include communication complexity, circuit complexity, and decision tree complexity.

Best, Worst and Average Case

In computer science, best, worst and average cases of a given algorithm express what the resourceusage is *at least*, *at most* and *on average*, respectively. Usually the resource being considered is running time, but it could also be memory or other resources.

In real-time computing, the worst-case execution time is often of particular concern since it is important to know how much time might be needed *in the worst case* to guarantee that the algorithm will always finish on time. Average performance and worst-case performance are the most used in algorithm analysis. Less widely found is best-case performance, but it does have uses: for example, where the best cases of individual tasks are known, they can be used to improve the accuracy of an overall worst-case analysis. Computer scientists use probabilistic analysis techniques, especially expected value, to determine expected running times.

The terms are used in other contexts; for example the worst- and best-case outcome of a planned-for epidemic, worst-case temperature to which an electronic circuit element is exposed, etc. Where components of specified tolerance are used, devices must be designed to work properly with the worst-case combination of tolerances and external conditions.

Best-case Performance

The term *best-case performance* is used in computer science to describe the way an algorithm behaves under optimal conditions. For example, the best case for a simple linear search on a list occurs when the desired element is the first element of the list.

Development and choice of algorithms is rarely based on best-case performance: most academic and commercial enterprises are more interested in improving average performance and worst-case performance. Algorithms may also be trivially modified to have good best-case running time by hard-coding solutions to a finite set of inputs, making the measure almost meaningless.

Worst-case Versus Average-case Performance

Worst-case performance analysis and average case performance analysis have some similarities, but in practice usually require different tools and approaches.

Determining what *average input* means is difficult, and often that average input has properties which make it difficult to characterise mathematically (consider, for instance, algorithms that are designed to operate on strings of text). Similarly, even when a sensible description of a particular "average case" (which will probably only be applicable for some uses of the algorithm) is possible, they tend to result in more difficult to analyse equations.

Worst-case analysis has similar problems: it is typically impossible to determine the exact worst-case scenario. Instead, a scenario is considered such that it is at least as bad as the worst case. For example, when analysing an algorithm, it may be possible to find the longest possible path through the algorithm (by considering the maximum number of loops, for instance) even if it is not possible to determine the exact input that would generate this path (indeed, such an input may not exist). This gives a *safe* analysis (the worst case is never underestimated), but one which is *pessimistic*, since there may be no input that would require this path.

Alternatively, a scenario which is thought to be close to (but not necessarily worse than) the real worst case may be considered. This may lead to an *optimistic* result, meaning that the analysis may actually underestimate the true worst case.

In some situations it may be necessary to use a pessimistic analysis in order to guarantee safety. Often however, a pessimistic analysis may be too pessimistic, so an analysis that gets closer to the real value but may be optimistic (perhaps with some known low probability of failure) can be a much more practical approach.

When analyzing algorithms which often take a small time to complete, but periodically require a much larger time, amortized analysis can be used to determine the worst-case running time over a (possibly infinite) series of operations. This amortized worst-case cost can be much closer to the average case cost, while still providing a guaranteed upper limit on the running time.

The worst-case analysis is related to the worst-case complexity.

Practical Consequences

Many problems with bad worst-case performance have good average-case performance. For problems we want to solve, this is a good thing: we can hope that the particular instances we care about are average. For cryptography, this is very bad: we want typical instances of a cryptographic problem to be hard. Here methods like random self-reducibility can be used for some specific problems to show that the worst case is no harder than the average case, or, equivalently, that the average case is no easier than the worst case.

On the other hand some algorithms like hash tables have very poor worst case behaviours, but a well written hash table of sufficient size will statistically never give the worst case; the average number of operations performed follows an exponential decay curve, and so the run time of an operation is statistically bounded.

Examples

- Linear search on a list of n elements. In the worst case, the search must visit every element once. This happens when the value being searched for is either the last element in the list, or is not in the list. However, on average, assuming the value searched for is in the list and each list element is equally likely to be the value searched for, the search visits only $n/2$ elements.

- Insertion sort applied to a list of n elements, assumed to be all different and initially in random order. On average, half the elements in a list $A_1 \dots A_j$ are less than element A_{j+1}, and half are greater. Therefore the algorithm compares the j+1-st element to be inserted on the average with half the already sorted sub-list, so $t_j = j/2$. Working out the resulting average-case running time yields a quadratic function of the input size, just like the worst-case running time.
- Quicksort applied to a list of n elements, again assumed to be all different and initially in random order. This popular sorting algorithm has an average-case performance of $O(n \log n)$, which contributes to making it a very fast algorithm in practice. But given a worst-case input, its performance degrades to $O(n^2)$.

Quicksort

Quicksort is a sorting algorithm developed byTony Hoare that, on average, makes $O(n\log n)$(big O notation) comparisons to sort n items. In the worst case, it makes $O(n^2)$ comparisons, though this behavior is rare. Quicksort is often faster in practice than other $O(n\log n)$algorithms. Additionally, quicksort's sequential and localized memory references work well with a cache. Quicksort can be implemented with anin-place partitioning algorithm, so the entire sort can be done with only $O(\log n)$ additional space.

Quicksort (also known as "partition-exchange sort") is a comparison sort and, in space efficient implementations, is not a stable sort.

History

The quicksort algorithm was developed in 1960 by Tony Hoare while in the Soviet Union, as a visiting student at Moscow State University. At that time, Hoare worked in a project on machine translation for the National Physical Laboratory. He developed the algorithm in order to sort the words to be translated, to make them more easily matched to an already-sorted Russian-to-English dictionary that was stored on magnetic tape.

Implementation Issues

Choice of Pivot

In very early versions of quicksort, the leftmost element of the partition would often be chosen as the pivot element. Unfortunately, this causes worst-case behavior on already sorted arrays, which is a rather common use-case.

The problem was easily solved by choosing either a random index for the pivot, choosing the middle index of the partition or (especially for longer partitions) choosing themedian of the first, middle and last element of the partition for the pivot (as recommended by R. Sedgewick).

Selecting a pivot element is also complicated by the existence of integer overflow.

If the boundary indices of the subarray being sorted are sufficiently large, the naïve expression for the middle index,*(left + right)/2*, will cause overflow and provide an invalid pivot index. This can be overcome by using, for example, *left + (right-left)/2* to index the middle element, at the cost of more complex arithmetic. Similar issues arise in some other methods of selecting the pivot element.

Optimizations

Two other important optimizations, also suggested by R. Sedgewick, as commonly acknowledged, and widely used in practice are:

- To make sure at most O(log N) space is used, recurse first into the smaller half of the array, and use a tail call to recurse into the other.
- Use insertion sort, which has a smaller constant factor and is thus faster on small arrays, for invocations on such small arrays (i.e. where the length is less than a threshold *t* determined experimentally). This can be implemented by leaving such arrays unsorted and running a singleinsertion sort pass at the end, because insertion sort handles nearly sorted arrays efficiently. A separate insertion sort of each small segment as they are identified adds the overhead of starting and stopping many small sorts, but avoids wasting effort comparing keys across the many segment boundaries, which keys will be in order due to the workings of the quicksort process. It also improves the cache use.

Parallelization

Like merge sort, quicksort can also be parallelized due to its divide-and-conquer nature. Individual in-place partition operations are difficult to parallelize, but once divided, different sections of the list can be sorted in parallel. The following is a straightforward approach: If we have p processors, we can divide a list of n elements into p sublists in $O(n)$ average time, then sort each of these in $O\left(\frac{n}{p}\log\frac{n}{p}\right)$ average time. Ignoring the $O(n)$ preprocessing and merge times, this is linear speedup. If the split is blind, ignoring the values, the merge naïvely costs $O(n)$. If the split partitions based on a succession of pivots, it is tricky to parallelize and naïvely costs $O(n)$. Given $O(\log n)$ or more processors, only $O(n)$ time is required overall, whereas an approach with linear speedup would achieve $O(\log n)$ time for overall.

One advantage of this simple parallel quicksort over other parallel sort algorithms is that no synchronization is required, but the disadvantage is that sorting is still $O(n)$ and only a sublinear speedup of $O(\log n)$ is achieved. A new thread is started as soon as a sublist is available for it to work on and it does not communicate with other threads. When all threads complete, the sort is done.

Other more sophisticated parallel sorting algorithms can achieve even better time bounds. For example, in 1991 David Powers described a parallelized quicksort (and a related radix sort) that can operate in $O(\log n)$ time on a CRCW PRAM with n processors by performing partitioning implicitly.

Formal Analysis

From the initial description it's not obvious that quicksort takes $O(n \log n)$ time on average. It's not hard to see that the partition operation, which simply loops over the elements of the array once, uses$O(n)$ time. In versions that perform concatenation, this operation is also $O(n)$.

In the best case, each time we perform a partition we divide the list into two nearly equal pieces. This means each recursive call processes a list of half the size. Consequently, we can make only log $n$nested calls before we reach a list of size 1. This means that the depth of the call tree is log n. But no two calls at the same level of the call tree process the same part of the original list; thus, each level of calls needs only $O(n)$ time all together (each call has some constant overhead, but since there are only $O(n)$ calls at each level, this is subsumed in the $O(n)$ factor). The result is that the algorithm uses only $O(n \log n)$ time.

An alternative approach is to set up a recurrence relation for the $T(n)$ factor, the time needed to sort a list of size n. Because a single quicksort call involves $O(n)$ factor work plus two recursive calls on lists of size $n / 2$ in the best case, the relation would be:

$$T(n) = O(n) + 2T\left(\frac{n}{2}\right).$$

The master theorem tells us that $T(n) = O(n \log n)$.

In fact, it's not necessary to divide the list this precisely; even if each pivot splits the elements with 99% on one side and 1% on the other (or any other fixed fraction), the call depth is still limited to100$logn$, so the total running time is still $O(n \log n)$.

In the worst case, however, the two sublists have size 1 and n " 1 (for example, if the array consists of the same element by value), and the call tree becomes a linear chain of n nested calls. The ith call does $O(n$ " $i)$ work, and $\sum_{i=0}^{n}(n-i) = O(n^2)$ The recurrence relation is:

$T(n) = O(n) + T(0) + T(n - 1) = O(n) + T(n - 1)$.

This is the same relation as for insertion sort and selection sort, and it solves to $T(n) = O(n^2)$. Given knowledge of which comparisons are performed by the sort, there are adaptive algorithms that are effective at generating worst-case input for quicksort on-the-fly, regardless of the pivot selection strategy.

Randomized Quicksort Expected Complexity

Randomized quicksort has the desirable property that, for any input, it requires only $O(n \log n)$expected time (averaged over all choices of pivots). But what makes random pivots a good choice? Suppose we sort the list and then divide it into four parts. The two parts in the middle will contain the best pivots; each of them is larger than at least 25% of the elements and smaller than at least 25% of the elements. If we could consistently choose an element from these two middle parts, we would only have to split the list at most 2log $_2n$ times before reaching lists of size 1, yielding an $O(n \log n)$ algorithm.

A random choice will only choose from these middle parts half the time. However, this is good enough. Imagine that you are flipping a coin over and over until you get k heads. Although this could take a long time, on average only $2k$ flips are required, and the chance that you won't get k heads after $100k$flips is highly improbable. By the same argument, quicksort's recursion will terminate on average at a call depth of only $2(2\log_2 n)$. But if its average call depth is $O(\log n)$, and each level of the call tree processes at most n elements, the total amount of work done on average is the product, $O(n \log n)$. Note that the algorithm does not have to verify that the pivot is in the middle half—if we hit it any constant fraction of the times, that is enough for the desired complexity.

The outline of a formal proof of the $O(n \log n)$ expected time complexity follows. Assume that there are no duplicates as duplicates could be handled with linear time pre- and post-processing, or considered cases easier than the analyzed. Choosing a pivot, uniformly at random from 0 to n-1, is then equivalent to choosing the size of one particular partition, uniformly at random from 0 to n-1. With this observation, the continuation of the proof is analogous to the one given in the average complexity section.

Average Complexity

Even if pivots aren't chosen randomly, quicksort still requires only $O(n \log n)$ time averaged over all possible permutations of its input. Because this average is simply the sum of the times over all permutations of the input divided by n factorial, it's equivalent to choosing a random permutation of the input. When we do this, the pivot choices are essentially random, leading to an algorithm with the same running time as randomized quicksort.

More precisely, the average number of comparisons over all permutations of the input sequence can be estimated accurately by solving the recurrence relation:

$$C(n) = n-1+\frac{1}{n}\sum_{i=0}^{n-1}(C(i)+C(n-i-1)) = 2n\ln n = 1.39n\log_2 n.$$

Here, n H 1 is the number of comparisons the partition uses. Since the pivot is equally likely to fall anywhere in the sorted list order, the sum is averaging over all possible splits.

This means that, on average, quicksort performs only about 39% worse than in its best case. In this sense it is closer to the best case than the worst case. Also note that a comparison sort cannot use less than $\log_2(n!)$ comparisons on average to sort n items (as explained in the article Comparison sort) and in case of large n, Stirling's approximation yields $\log_2(n!) \approx n(\log_2 n - \log_2 e)$, so quicksort is not much worse than an ideal comparison sort. This fast average runtime is another reason for quicksort's practical dominance over other sorting algorithms.

Space Complexity

The space used by quicksort depends on the version used.

Quicksort has a space complexity of $O(\log n)$, even in the worst case, when it is carefully implemented ensuring the following two properties:

- in-place partitioning is used. This requires $O(1)$ space.
- After partitioning, the partition with the fewest elements is (recursively) sorted first, requiring at most $O(\log n)$ space. Then the other partition is sorted using tail recursion or iteration, which doesn't add to the call stack. This idea, as discussed above, was described by R. Sedgewick.

The version of quicksort with in-place partitioning uses only constant additional space before making any recursive call. However, if it has made $O(\log n)$ nested recursive calls, it needs to store a constant amount of information from each of them. Since the best case makes at most $O(\log n)$ nested recursive calls, it uses $O(\log n)$ space. The worst case makes $O(n)$ nested recursive calls, and so needs $O(n)$ space; Sedgewick's improved version using tail recursion requires $O(\log n)$ space in the worst case.

We are eliding a small detail here, however. If we consider sorting arbitrarily large lists, we have to keep in mind that our variables like *left* and *right* can no longer be considered to occupy constant space; it takes $O(\log n)$ bits to index into a list of n items. Because we have variables like this in every stack frame, in reality quicksort requires $O((\log n)^2)$ bits of space in the best and average case and$O(n \log n)$ space in the worst case. This isn't too terrible, though, since if the list contains mostly distinct elements, the list itself will also occupy $O(n \log n)$ bits of space.

The not-in-place version of quicksort uses $O(n)$ space before it even makes any recursive calls. In the best case its space is still limited to $O(n)$, because each level of the recursion uses half as much space as the last, and

$$\sum_{i=0}^{\infty} \frac{n}{2^i} = 2n.$$

Its worst case is dismal, requiring

$$\sum_{i=0}^{n}(n-i+1)=O(n^2)$$

space, far more than the list itself. If the list elements are not themselves constant size, the problem grows even larger; for example, if most of the list elements are distinct, each would require about $O(\log n)$ bits, leading to a best-case $O(n \log n)$ and worst-case $O(n^2 \log n)$ space requirement.

Selection-based Pivoting

A selection algorithm chooses the kth smallest of a list of numbers; this is an easier problem in general than sorting. One simple but effective selection algorithm works nearly in the same manner as quicksort, except that instead of making recursive calls on both sublists, it only makes a single tail-recursive call on the sublist which contains the desired element. This small change lowers the average complexity to linear or $O(n)$ time, and makes it an in-place algorithm. A variation on this algorithm brings the worst-case time down to $O(n)$ (see selection algorithm for more information).

Conversely, once we know a worst-case $O(n)$ selection algorithm is available, we can use it to find the ideal pivot (the median) at every step of quicksort, producing a variant with worst-case $O(n \log n)$ running time. In practical implementations, however, this variant is considerably slower on average.

Another variant is to choose the Median of Medians as the pivot element instead of the median itself for partitioning the elements. While maintaining the asymptotically optimal run time complexity of $O(n\log n)$ (by preventing worst case partitions), it is also considerably faster than the variant that chooses the median as pivot.

Variants

There are four well known variants of quicksort:

- Balanced quicksort: choose a pivot likely to represent the middle of the values to be sorted, and then follow the regular quicksort algorithm.
- External quicksort: The same as regular quicksort except the pivot is replaced by a buffer. First, read the M/2 first and last elements into the buffer and sort them. Read the next element from the beginning or end to balance writing. If the next

element is less than the least of the buffer, write it to available space at the beginning. If greater than the greatest, write it to the end. Otherwise write the greatest or least of the buffer, and put the next element in the buffer. Keep the maximum lower and minimum upper keys written to avoid resorting middle elements that are in order. When done, write the buffer. Recursively sort the smaller partition, and loop to sort the remaining partition. This is a kind of three-way quicksort in which the middle partition (buffer) represents a sorted subarray of elements that are *approximately* equal to the pivot.

- Three-way radix quicksort (developed by Sedgewick and also known as multikey quicksort): is a combination of radix sort and quicksort. Pick an element from the array (the pivot) and consider the first character (key) of the string (multikey). Partition the remaining elements into three sets: those whose corresponding character is less than, equal to, and greater than the pivot's character. Recursively sort the "less than" and "greater than" partitions on the same character. Recursively sort the "equal to" partition by the next character (key). Given we sort using bytes or words of length W bits, the best case is O(KN) and the worst case $O(2^K N)$ or at least $O(N^2)$ as for standard quicksort, given for unique keys $N<2^K$, and K is a hidden constant in all standard comparison sortalgorithms including quicksort. This is a kind of three-way quicksort in which the middle partition represents a (trivially) sorted subarray of elements that are *exactly* equal to the pivot.
- Quick radix sort (also developed by Powers as a o(K) parallel PRAM algorithm). This is again a combination of radix sort and quicksort but the quicksort left/right partition decision is made on successive bits of the key, and is thus O(KN) for N K-bit keys. Note that all comparison sortalgorithms effectively assume an ideal K of O(logN) as if k is smaller we can sort in O(N) using a hash table or integer sorting, and if K >> logN but elements are unique within O(logN) bits, the remaining bits will not be looked at by either quicksort or quick radix sort, and otherwise all comparison sorting algorithms will also have the same overhead of looking through O(K) relatively useless bits but quick radix sort will avoid the worst case $O(N^2)$ behaviours of standard quicksort and quick radix sort, and will

be faster even in the best case of those comparison algorithms under these conditions of uniqueprefix(K) >> logN. See Powers for further discussion of the hidden overheads in comparison, radix and parallel sorting.

Comparison with Other Sorting Algorithms

Quicksort is a space-optimized version of the binary tree sort. Instead of inserting items sequentially into an explicit tree, quicksort organizes them concurrently into a tree that is implied by the recursive calls. The algorithms make exactly the same comparisons, but in a different order. An often desirable property of a sorting algorithm is stability - that is the order of elements that compare equal is not changed, allowing controlling order of multikey tables (e.g. directory or folder listings) in a natural way. This property is hard to maintain for in situ (or in place) quicksort (that uses only constant additional space for pointers and buffers, and logN additional space for the management of explicit or implicit recursion). For variant quicksorts involving extra memory due to representations using pointers (e.g. lists or trees) or files (effectively lists), it is trivial to maintain stability. The more complex, or disk-bound, data structures tend to increase time cost, in general making increasing use of virtual memory or disk.

The most direct competitor of quicksort is heapsort. Heapsort's worst-case running time is always $O(n\log n)$. But, heapsort is assumed to be on average somewhat slower than standard in-place quicksort. This is still debated and in research, with some publications indicating the opposite. Introsort is a variant of quicksort that switches to heapsort when a bad case is detected to avoid quicksort's worst-case running time. If it is known in advance that heapsort is going to be necessary, using it directly will be faster than waiting for introsort to switch to it.

Quicksort also competes with mergesort, another recursive sort algorithm but with the benefit of worst-case $O(n \log n)$ running time. Mergesort is a stable sort, unlike standard in-place quicksort and heapsort, and can be easily adapted to operate on linked lists and very large lists stored on slow-to-access media such as disk storage or network attached storage. Although quicksort can be written to operate on linked lists, it will often suffer from poor pivot choices without random access. The main disadvantage of mergesort is that, when operating on arrays, it requires $O(n)$ auxiliary space in the best case, whereas the variant of quicksort with in-place partitioning and tail recursion uses only $O(\log n)$ space. (Note that when operating on linked lists, mergesort only requires a small, constant amount of auxiliary storage.)

Bucket sort with two buckets is very similar to quicksort; the pivot in this case is effectively the value in the middle of the value range, which does well on average for uniformly distributed inputs.

The best, worst and average case complexity refer to three different ways of measuring the time complexity (or any other complexity measure) of different inputs of the same size. Since some inputs of size n may be faster to solve than others, we define the following complexities:

- Best-case complexity: This is the complexity of solving the problem for the best input of size n.
- Worst-case complexity: This is the complexity of solving the problem for the worst input of size n.
- Average-case complexity: This is the complexity of solving the problem on an average. This complexity is only defined with respect to a probability distribution over the inputs. For instance, if all inputs of the same size are assumed to be equally likely to appear, the average case complexity can be defined with respect to the uniform distribution over all inputs of size n.

For example, consider the dc sorting algorithm quicksort. This solves the problem of sorting a list of integers that is given as the input. The worst-case is when the input is sorted or sorted in reverse order, and the algorithm takes time $O(n^2)$ for this case. If we assume that all possible permutations of the input list are equally likely, the average time taken for sorting is $O(n \log n)$. The best case occurs when each pivoting divides the list in half, also needing $O(n \log n)$ time.

Upper and Lower Bounds on the Complexity of Problems

To classify the computation time (or similar resources, such as space consumption), one is interested in proving upper and lower bounds on the minimum amount of time required by the most efficient algorithm solving a given problem. The complexity of an algorithm is usually taken to be its worst-case complexity, unless specified otherwise. Analyzing a particular algorithm falls under the field of analysis of algorithms. To show an upper bound $T(n)$ on the time complexity of a problem, one needs to show only that there is a particular algorithm with running time at most $T(n)$. However, proving lower bounds is much more difficult, since lower bounds make a statement about all possible algorithms that solve a given problem. The phrase "all possible algorithms" includes not just the algorithms known today, but any algorithm that might be discovered in the future. To show a lower bound of $T(n)$ for a problem requires showing that no algorithm can have time complexity lower than $T(n)$.

Upper and lower bounds are usually stated using the big Oh notation, which hides constant factors and smaller terms. This makes the bounds independent of the specific details of the computational model used. For instance, if $T(n) = 7n^2 + 15n + 40$, in big Oh notation one would write $T(n) = O(n^2)$.

Big O notation

In mathematics, big O notation is used to describe the limiting behavior of a function when the argument tends towards a particular value or infinity, usually in terms of simpler functions. It is a member of a larger family of notations that is called Landau notation, Bachmann-Landau notation, or asymptotic notation. In computer science, big O notation is used to classify algorithms by how they respond (*i.e.*, in their processing time or working space requirements) to changes in input size.

Big O notation characterizes functions according to their growth rates: different functions with the same growth rate may be represented using the same O notation. A description of a function in terms of big O notation usually only provides an upper bound on the growth rate of the function. Associated with big O notation are several related notations, using the symbols o, Ω, ω, and È, to describe other kinds of bounds on asymptotic growth rates. Big O notation is also used in many other fields to provide similar estimates.

Formal Definition

Let $f(x)$ and $g(x)$ be two functions defined on some subset of the real numbers. One writes

$$f(x) = O(g(x)) \text{ as } x \to \infty$$

if and only if, there is a sufficiently large constant M such that for all sufficiently large values of x, $f(x)$ is at most M multiplied by $g(x)$ in absolute value. That is, $f(x) = O(g(x))$ if and only if there exists a positive real number M and a real number x_0 such that

$$| f(x) | \le M | g(x) | \text{ for all } x > x_0.$$

In many contexts, the assumption that we are interested in the growth rate as the variable x goes to infinity is left unstated, and one writes more simply that $f(x) = O(g(x))$. The notation can also be used to describe the behavior of f near some real number a (often, $a = 0$): we say

$$f(x) = O(g(x)) \text{ as } x \to a$$

if and only if there exist positive numbers δ and M such that

$$| f(x) | \le M | g(x) | \text{ for } | x - a | < \delta.$$

If $g(x)$ is non-zero for values of x sufficiently close to a, both of these definitions can be unified using the limit superior:

$$f(x) = O(g(x)) \text{ as } x \to a$$

if and only if

$$\limsup_{x \to a} \left| \frac{f(x)}{g(x)} \right| < \infty.$$

Example

In typical usage, the formal definition of O notation is not used directly; rather, the O notation for a function $f(x)$ is derived by the following simplification rules:

- If $f(x)$ is a sum of several terms, the one with the largest growth rate is kept, and all others omitted.
- If $f(x)$ is a product of several factors, any constants (terms in the product that do not depend on x) are omitted.

For example, let $f(x) = 6x^4$ ” $2x^3 + 5$, and suppose we wish to simplify this function, using O notation, to describe its growth rate as x approaches infinity. This function is the sum of three terms: $6x^4$, “$2x^3$, and 5. Of these three terms, the one with the highest growth rate is the one with the largest exponent as a function of x, namely $6x^4$. Now one may apply the second rule: $6x^4$ is a product of 6 and x^4 in which the first factor does not depend on x. Omitting this factor results in the simplified form x^4. Thus, we say that $f(x)$ is a big-oh of (x^4) or mathematically we can write $f(x) = O(x^4)$. One may confirm this calculation using the formal definition: let $f(x) = 6x^4$ ” $2x^3 + 5$ and $g(x) = x^4$. Applying the formal definition from above, the statement that $f(x) = O(x^4)$ is equivalent to its expansion,

$$| f(x) | \le M | g(x) |$$

for some suitable choice of x_0 and M and for all $x > x_0$. To prove this, let $x_0 = 1$ and $M = 13$. Then, for all $x > x_0$:

$$\begin{aligned} |6x^4 - 2x^3 + 5| &\le 6x^4 + |2x^3| + 5 \\ &\le 6x^4 + 2x^4 + 5x^4 \\ &\le 13x^4 \\ &\le 13|x^4| \end{aligned}$$

so

$$|6x^4 - 2x^3 + 5| \le 13|x^4|.$$

Usage

Big O notation has two main areas of application. In mathematics, it is commonly used to describe how closely a finite series approximates a given function, especially in the case of a truncated Taylor series or asymptotic expansion. In computer science, it is useful in the analysis of algorithms. In both applications, the function $g(x)$ appearing within the $O(...)$ is typically chosen to be as simple as possible, omitting constant factors and lower order terms. There are two formally close, but noticeably different, usages of this notation: infinite asymptotics and infinitesimal asymptotics. This distinction is only in application and not in principle, however—the formal definition for the "big O" is the same for both cases, only with different limits for the function argument.

Infinite Asymptotics

Big O notation is useful when analyzing algorithms for efficiency. For example, the time (or the number of steps) it takes to complete a problem of size n might be found to be $T(n) = 4n^2$ " $2n + 2$. As $n$grows large, the n^2 term will come to dominate, so that all other terms can be neglected — for instance when $n = 500$, the term $4n^2$ is 1000 times as large as the $2n$ term.

Ignoring the latter would have negligible effect on the expression's value for most purposes. Further, the coefficients become irrelevant if we compare to any other order of expression, such as an expression containing a term n^3or n^4. Even if $T(n) = 1{,}000{,}000n^2$, if $U(n) = n^3$, the latter will always exceed the former once n grows larger than 1,000,000 ($T(1{,}000{,}000) = 1{,}000{,}000^3 = U(1{,}000{,}000)$). Additionally, the number of steps depends on the details of the machine model on which the algorithm runs, but different types of machines typically vary by only a constant factor in the number of steps needed to execute an algorithm. So the big O notation captures what remains: we write either

$$T(n) = O(n^2)$$

or

$$T(n) \in O(n^2)$$

and say that the algorithm has *order of* n^2 time complexity. Note that "=" is not meant to express "is equal to" in its normal mathematical sense, but rather a more colloquial "is", so the second expression is technically accurate (see the "Equals sign" discussion below) while the first is a common abuse of notation.

Infinitesimal Asymptotics

Big O can also be used to describe the error term in an approximation to a mathematical function. The most significant terms are written explicitly, and then the least-significant terms are summarized in a single big O term. For example,

$$e^x = 1 + x + \frac{x^2}{2} + O(x^3) \qquad \text{as } x \to 0$$

expresses the fact that the error, the difference $e^x - (1 + x + x^2 / 2)$ is smaller in absolute value than some constant times $| x^3 |$ when x is close enough to 0.

Properties

If a function $f(n)$ can be written as a finite sum of other functions, then the fastest growing one determines the order of $f(n)$. For example

$$f(n) = 9 \log n + 5(\log n)^3 + 3n^2 + 2n^3 = O(n^3)$$

In particular, if a function may be bounded by a polynomial in n, then as n tends to *infinity*, one may disregard *lower-order* terms of the polynomial. $O(n^c)$ and $O(c^n)$ are very different. The latter grows much, much faster, no matter how big the constant c is (as long as it is greater than one). A function that grows faster than any power of n is called *superpolynomial*. One that grows more slowly than any exponential function of the form c^n is called *subexponential*. An algorithm can require time that is both superpolynomial and subexponential; examples of this include the fastest known algorithms for integer factorization. $O(\log n)$ is exactly the same as $O(\log(n^c))$. The logarithms differ only by a constant factor (since $\log(n^c) = c \log n$) and thus the big O notation ignores that. Similarly, logs with different constant bases are equivalent. Exponentials with different bases, on the other hand, are not of the same order. For example, 2^n and 3^n are not of the same order. Changing units may or may not affect the order of the resulting algorithm.

Changing units is equivalent to multiplying the appropriate variable by a constant wherever it appears. For example, if an algorithm runs in the order of n^2, replacing n bycn means the algorithm runs in the order of c^2n^2, and the big O notation ignores the constant c^2. This can be written as $c^2n^2 \in O(n^2)$. If, however, an algorithm runs in the order of 2^n, replacing n with cn gives $2^{cn} = (2^c)^n$. This is not equivalent to 2^n in general. Changing of variable may affect the order of the resulting algorithm. For example, if an algorithm's running time is $O(n)$ when measured in terms of the number n of *digits* of an input number x, then its running time is $O(\log x)$ when measured as a function of the input number x itself, because n = È($\log x$).

Product

$$f_1 \in O(g_1) \text{ and } f_2 \in O(g_2) \Rightarrow f_1 f_2 \in O(g_1 g_2)$$

$$f \cdot O(g) \subset O(fg)$$

Sum

$$f_1 \in O(g_1) \text{ and } f_2 \in O(g_2) \Rightarrow f_1 + f_2 \in O(|g_1| + |g_2|)$$

This implies $f_1 \in O(g)$ and $f_2 \in O(g) \Rightarrow f_1 + f_2 \in O(g)$ which means that$O(g)$ is a convex cone.

If f and g are positive functions, $f + O(g) \in O(f + g)$

Multiplication by a Constant

Let k be a constant. Then:

$O(kg) = O(g)$ if k is nonzero.

$f \in O(g) \Rightarrow kf \in O(g)$.

Multiple Variables

Big O (and little o, and Ω...) can also be used with multiple variables. To define Big O formally for multiple variables, suppose $f(\vec{x})$ and $g(\vec{x})$ are two functions defined on some subset of $\mathbb{R}^n$ We say

$$f(\vec{x}) \text{ is } O(g(\vec{x})) \text{ as } \vec{x} \to \infty$$

if and only if

$\exists C \exists M > 0$ such that $|f(\vec{x})| \leq C|g(\vec{x})|$ for all $\vec{x}$ with $x_i > M$ for all i.

For example, the statement

$$f(n,m) = n^2 + m^3 + O(n+m) \text{ as } n,m \to \infty$$

asserts that there exist constants C and M such that

$$\forall n,m > M : | g(n,m) | \leq C(n+m),$$

where $g(n,m)$ is defined by

$$f(n,m) = n^2 + m^3 + g(n,m).$$

Note that this definition allows all of the coordinates of $\vec{x}$ to increase to infinity. In particular, the statement

$$f(n,m) = O(n^m) \text{ as } n,m \to \infty$$

(i.e., $\exists C \exists M \forall n \forall m \ldots$) is quite different from

$$\forall m : f(n,m) = O(n^m) \text{ as } n \to \infty$$

(i.e., $\forall m \exists C \exists M \forall n \ldots$).

Matters of Notation

Equals Sign

The statement "$f(x)$ is $O(g(x))$" as defined above is usually written as $f(x) = O(g(x))$. Some consider this to be an abuse of notation, since the use of the equals sign could be misleading as it suggests a symmetry that this statement does not have. As de Bruijn says, $O(x) = O(x^2)$ is true but $O(x^2) = O(x)$ is not. Knuth describes such statements as "one-way equalities", since if the sides could be reversed, "we could deduce ridiculous things like $n = n^2$ from the identities $n = O(n^2)$ and$n^2 = O(n^2)$."

For these reasons, it would be more precise to use set notation and write $f(x)$ " $O(g(x))$, thinking of $O(g(x))$ as the class of all functions $h(x)$ such that $|h(x)| \leq C|g(x)|$ for some constant C.However, the use of the equals sign is customary. Knuth pointed out that "mathematicians customarily use the = sign as they use the word 'is' in English: Aristotle is a man, but a man isn't necessarily Aristotle."

Other Arithmetic Operators

Big O notation can also be used in conjunction with other arithmetic operators in more complicated equations. For example, $h(x) + O(f(x))$ denotes the collection of functions having the growth of $h(x)$ plus a part whose growth is limited to that of $f(x)$. Thus,

$$g(x) = h(x) + O(f(x))$$

expresses the same as

$$g(x) - h(x) \in O(f(x)).$$

Example

Suppose an algorithm is being developed to operate on a set of n elements. Its developers are interested in finding a function $T(n)$ that will express how long the algorithm will take to run (in some arbitrary measurement of time) in terms of the number of elements in the input set. The algorithm works by first calling a subroutine to sort the elements in the set and then perform its own operations. The sort has a known time complexity of $O(n^2)$, and after the subroutine runs the algorithm must take an additional $55n^3 + 2n + 10$ time before it terminates. Thus the overall time complexity of the algorithm can be expressed as

$$T(n) = O(n^2) + 55n^3 + 2n + 10.$$

This can perhaps be most easily read by replacing $O(n^2)$ with "some function that grows asymptotically slower than n^2". Again, this usage disregards some of the formal meaning of the "=" and "+" symbols, but it does allow one to use the big O notation as a kind of convenient placeholder.

Declaration of Variables

Another feature of the notation, although less exceptional, is that function arguments may need to be inferred from the context when several variables are involved. The following two right-hand side big O notations have dramatically different meanings:

$$f(m) = O(m^n),$$

$$g(n) = O(m^n).$$

The first case states that $f(m)$ exhibits polynomial growth, while the second, assuming $m > 1$, states that $g(n)$ exhibits exponential growth. To avoid confusion, some authors use the notation

$$g(x) \in O(f(x)).$$

rather than the less explicit

$$g \in O(f),$$

Multiple Usages

In more complicated usage, $O(...)$ can appear in different places in an equation, even several times on each side. For example, the following are true for $n \to \infty$

$$(n+1)^2 = n^2 + O(n)$$
$$(n + O(n^{1/2}))(n + O(\log n))^2 = n^3 + O(n^{5/2})$$
$$n^{O(1)} = O(e^n).$$

The meaning of such statements is as follows: for *any* functions which satisfy each $O(...)$ on the left side, there are *some* functions satisfying each $O(...)$ on the right side, such that substituting all these functions into the equation makes the two sides equal. For example, the third equation above means: "For any function $f(n) = O(1)$, there is some function $g(n) = O(e^n)$ such that $n^{f(n)} = g(n)$." In terms of the "set notation" above, the meaning is that the class of functions represented by the left side is a subset of the class of functions represented by the right side. In this use the "=" is a formal symbol that unlike the usual use of "=" is not a symmetric relation. Thus for example $n^{O(1)} = O(e^n)$ does not imply the false statement $O(e^n) = n^{O(1)}$.

Related Asymptotic Notations

Big O is the most commonly used asymptotic notation for comparing functions, although in many cases Big O may be replaced with Big Theta È for asymptotically tighter bounds. Here, we define some related notations in terms of Big O, progressing up to the family of Bachmann–Landau notations to which Big O notation belongs.

Little-o Notation

The relation $f(x) \in o(g(x))$ is read as "$f(x)$ is little-o of $g(x)$". Intuitively, it means that $g(x)$grows much faster than $f(x)$, or similarly, the growth of $f(x)$ is nothing compared to that of $g(x)$. It assumes that f and g are both functions of one variable. Formally, $f(n) = o(g(n))$ as $n \rightarrow$ " means that for every positive constant ε there exists a constant N such that

$$|f(n)| \leq \epsilon |g(n)| \qquad \text{for all } n \geq N .$$

Note the difference between the earlier formal definition for the big-O notation, and the present definition of little-o: while the former has to be true for *at least one* constant M the latter must hold for*every* positive constant ε, however small. In this way little-o notation makes a stronger statement than the corresponding big-O notation: every function that is little-o of g is also big-O of g, but not every function that is big-O g is also little-o of g (for instance g itself is not, unless it is identically zero near "). If $g(x)$ is nonzero, or at least becomes nonzero beyond a certain point, the relation $f(x) = o(g(x))$ is equivalent to

$$\lim_{x\to\infty}\frac{f(x)}{g(x)}=0.$$

For example,

- $2x \in o(x^2)$
- $2x^2 \notin o(x^2)$
- $1/x \in o(1)$

Little-o notation is common in mathematics but rarer in computer science. In computer science the variable (and function value) is most often a natural number. In mathematics, the variable and function values are often real numbers. The following properties can be useful:

- $o(f)+o(f) \subseteq o(f)$
- $o(f)o(g) \subseteq o(fg)$
- $o(o(f)) \subseteq o(f)$
- $o(f) \subset O(f)$ (and thus the above properties apply with most combinations of o and O).

As with big O notation, the statement "*f*(*x*) is *o*(*g*(*x*))" is usually written as *f*(*x*) = *o*(*g*(*x*)), which is a slight abuse of notation.

Family of Bachmann–Landau Notations

Bachmann–Landau notation was designed around several mnemonics, as shown in the *As* $n \to \infty$*, eventually...* column above and in the bullets below. To conceptually access these mnemonics, "omicron" can be read "o-*micron*" and "omega" can be read "o-*mega*". Also, the lower-case versus capitalization of the Greek letters in Bachmann–Landau notation is mnemonic.

- The *o-micron mnemonic*: The o-*micron* reading of $f(n) \in O(g(n))$ and of $f(n) \in o(g(n))$ can be thought of as "O-*smaller than*" and "o-*smaller than*", respectively. This *micro*/smaller mnemonic refers to: for sufficiently large input parameter(s), *f* grows at a rate that may henceforth be less than *cg* regarding $g \in O(f)$ or $g \in o(f)$
- The *o-mega mnemonic*: The o-*mega* reading of $f(n) \in \Omega(g(n))$ and of $f(n) \in \omega(g(n))$ can be thought of as "O-*larger than*". This *mega*/larger mnemonic refers to: for sufficiently large input parameter(s), *f* grows at a rate that may henceforth be greater than *cg* regarding $g \in \Omega(f)$ or $g \in \omega(f)$.

- The *upper-case mnemonic*: This mnemonic reminds us when to use the upper-case Greek letters in $f(n) \in O(g(n))$ and $f(n) \in \Omega(g(n))$ for sufficiently large input parameter(s), f grows at a rate that may henceforth be equal to cg regarding $g \in O(f)$.
- The *lower-case mnemonic*: This mnemonic reminds us when to use the lower-case Greek letters in $f(n) \in o(g(n))$ and $f(n) \in \omega(g(n))$ for sufficiently large input parameter(s), f grows at a rate that is henceforth inequal to cg regarding .

Aside from Big O notation, the Big Theta È and Big Omega Ω notations are the two most often used in computer science; the Small Omega ω notation is rarely used in computer science.

Use in Computer Science

Informally, especially in computer science, the Big O notation often is permitted to be somewhat abused to describe an asymptotic tight bound where using Big Theta È notation might be more factually appropriate in a given context.

For example, when considering a function $T(n) = 73n^3 + 22n^2 + 58$, all of the following are generally acceptable, but tightnesses of bound (i.e., bullets 2 and 3 below) are usually strongly preferred over laxness of bound (i.e., number 1 below).

1. $T(n) = O(n^{100})$, which is identical to $T(n)$ " $O(n^{100})$
2. $T(n) = O(n^3)$, which is identical to $T(n)$ " $O(n^3)$
3. $T(n) =$ È(n^3), which is identical to $T(n)$ " È(n^3).

The equivalent English statements are respectively:

1. $T(n)$ grows asymptotically no faster than n^{100}
2. $T(n)$ grows asymptotically no faster than n^3
3. $T(n)$ grows asymptotically as fast as n^3.

So while all three statements are true, progressively more information is contained in each. In some fields, however, the Big O notation (number 2 in the lists above) would be used more commonly than the Big Theta notation (bullets number 3 in the lists above) because functions that grow more slowly are more desirable. For example, if $T(n)$ represents the running time of a newly developed algorithm for input size n, the inventors and users of the algorithm might be more inclined to put an upper asymptotic bound on how long it will take to run without making an explicit statement about the lower asymptotic bound.

Extensions to the Bachmann–Landau Notations

Another notation sometimes used in computer science is ϕ (read *soft-O*): $f(n) = \phi(g(n))$ is shorthand for $f(n) = O(g(n) \log^k g(n))$ for some k. Essentially, it is Big O notation, ignoring logarithmic factors because the growth-rate effects of some other super-logarithmic function indicate a growth-rate explosion for large-sized input parameters that is more important to predicting bad run-time performance than the finer-point effects contributed by the logarithmic-growth factor(s). This notation is often used to obviate the "nitpicking" within growth-rates that are stated as too tightly bounded for the matters at hand (since $\log^k n$ is always $o(n^\varepsilon)$ for any constant k and any $\varepsilon > 0$). The L notation, defined as

$$L_n[\alpha,c] = O\left(e^{(c+o(1))(\ln n)^\alpha (\ln\ln n)^{1-\alpha}}\right),$$

is convenient for functions that are between polynomial and exponential.

Generalizations and Related Usages

The generalization to functions taking values in any normed vector space is straightforward (replacing absolute values by norms), where f and g need not take their values in the same space. A generalization to functions g taking values in any topological group is also possible. The "limiting process" $x \to x_o$ can also be generalized by introducing an arbitrary filter base, i.e. to directed nets $f$and g. The o notation can be used to define derivatives and differentiability in quite general spaces, and also (asymptotical)

equivalence of functions,

$$f \sim g \Leftrightarrow (f - g) \in o(g)$$

which is an equivalence relation and a more restrictive notion than the relationship "f is È(g)" from above. (It reduces to lim f / g = 1 if f and g are positive real valued functions.) For example, $2x$ is È(x), but $2x$ " x is not $o(x)$.

Graph Theory

It is often useful to bound the running time of graph algorithms. Unlike most other computational problems, for a graph $G = (V, E)$ there are two relevant parameters describing the size of the input: the number $|V|$ of vertices in the graph and the number $|E|$ of edges in the graph. Inside asymptoticnotation (and only there), it is common to use the symbols $V$and E, when someone really means $|V|$ and $|E|$.

This convention simplifies asymptotic functions and make them easily readable. The symbols$V$ and E are never used inside asymptotic notation with their literal meaning, since the number of vertices and edges must be non-negative, so this abuse of notation does not risk ambiguity. For example $O(E + V\log V)$ means $O((V,E) \mapsto |E| + |V| \cdot \log |V|)$ for a suitable metric of graphs. Another common convention—referring to the values $|V|$ and $|E|$ by the names n and m, respectively—sidesteps this ambiguity.

History

The notation was first introduced by number theorist Paul Bachmann in 1894, in the second volume of his book *Analytische Zahlentheorie* ("analytic number theory"), the first volume of which (not yet containing big O notation) was published in 1892. The notation was popularized in the work of number theorist Edmund Landau; hence it is sometimes called a Landau symbol. It was popularized in computer science by Donald Knuth, who re-introduced the related Omega and Theta notations. He also noted that the (then obscure) Omega notation had been introduced by Hardy and Littlewoodunder a slightly different meaning, and proposed the current definition. Hardy's symbols were (in terms of the modern O notation)

$$f \lesssim g \Leftrightarrow f \in O(g) \text{ and } f \ll g \Leftrightarrow f \in o(g);$$

other similar symbols were sometimes used, such as $\preceq$ and $\prec\!\!\prec$. The big-O, standing for "order of", was originally a capital omicron; today the identical-looking Latin capital letter O is used, but never the digit zero.

Complexity Classes

Defining Complexity Classes

A complexity class is a set of problems of related complexity. Simpler complexity classes are defined by the following factors:

- The type of computational problem: The most commonly used problems are decision problems. However, complexity classes can be defined based on function problems, counting problems,optimization problems, promise problems, etc.
- The model of computation: The most common model of computation is the deterministic Turing machine, but many complexity classes are based on nondeterministic Turing machines, Boolean circuits, quantum Turing machines, monotone circuits, etc.

- The resource (or resources) that are being bounded and the bounds: These two properties are usually stated together, such as "polynomial time", "logarithmic space", "constant depth", etc.

Of course, some complexity classes have complex definitions that do not fit into this framework. Thus, a typical complexity class has a definition like the following:

The set of decision problems solvable by a deterministic Turing machine within time $f(n)$. (This complexity class is known as DTIME($f(n)$).)

But bounding the computation time above by some concrete function $f(n)$ often yields complexity classes that depend on the chosen machine model. For instance, the language $\{xx \mid x$ is any binary string$\}$ can be solved in linear time on a multi-tape Turing machine, but necessarily requires quadratic time in the model of single-tape Turing machines. If we allow polynomial variations in running time,Cobham-Edmonds thesis states that "the time complexities in any two reasonable and general models of computation are polynomially related" (Goldreich 2008, Chapter 1.2). This forms the basis for the complexity class P, which is the set of decision problems solvable by a deterministic Turing machine within polynomial time. The corresponding set of function problems is FP.

Important Complexity Classes

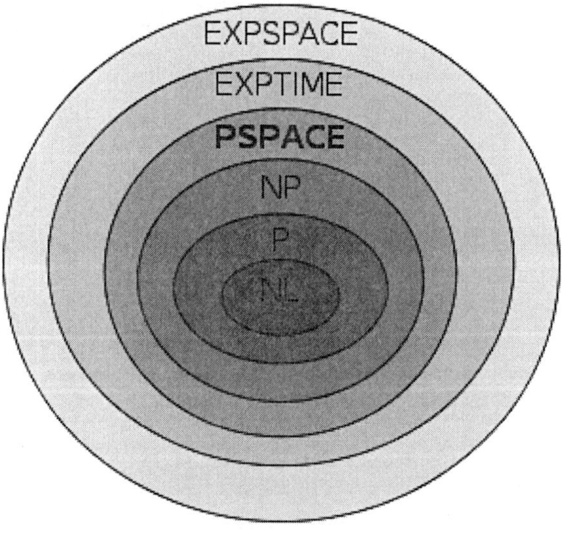

Figure: A representation of the relation among complexity classes

Many important complexity classes can be defined by bounding the time or space used by the algorithm. Some important complexity classes of decision problems defined in this manner are the following:

Complexity class	Model of computation	Resource constraint
DTIME($f(n)$)	Deterministic Turing machine	Time $f(n)$
P	Deterministic Turing machine	Time poly(n)
EXPTIME	Deterministic Turing machine	Time $2^{\text{poly}(n)}$
NTIME($f(n)$)	Non-deterministic Turing machine	Time $f(n)$
NP	Non-deterministic Turing machine	Time poly(n)
NEXPTIME	Non-deterministic Turing machine	Time $2^{\text{poly}(n)}$
DSPACE($f(n)$)	Deterministic Turing machine	Space $f(n)$
L	Deterministic Turing machine	Space O(log n)
PSPACE	Deterministic Turing machine	Space poly(n)
EXPSPACE	Deterministic Turing machine	Space $2^{\text{poly}(n)}$
NSPACE($f(n)$)	Non-deterministic Turing machine	Space $f(n)$
NL	Non-deterministic Turing machine	Space O(log n)
NPSPACE	Non-deterministic Turing machine	Space poly(n)
NEXPSPACE	Non-deterministic Turing machine	Space $2^{\text{poly}(n)}$

It turns out that PSPACE = NPSPACE and EXPSPACE = NEXPSPACE by Savitch's theorem.

Other important complexity classes include BPP, ZPP and RP, which are defined using probabilistic Turing machines; AC and NC, which are defined using Boolean circuits and BQP and QMA, which are defined using quantum Turing machines. #P is an important complexity class of counting problems (not decision problems). Classes like IP and AM are defined using Interactive proof systems. ALL is the class of all decision problems.

Boolean Circuit

A Boolean circuit is a mathematical model of computation used in studying computational complexity theory. Boolean circuits are the main object of study in circuit complexity and are a special kind of circuits; a formal language can be decided by a family of Boolean circuits, one circuit for each possible input length. In addition, they are used as a formal model for combinational logic indigital electronics.

Boolean circuits are defined in terms of the gates they contain. For example, a circuit might containbinary AND and OR gates and unary NOT gates, or be entirely described by binary NAND gates. Each gate corresponds to some Boolean function that takes k bits as input and outputs a single bit. Several important complexity measures can be defined on Boolean circuits, including circuit depth, circuit size, and number of alternations. In a circuit family, we consider the size complexity of a family, for example, to be the function of n that gives the size of the circuit that decides inputs of length n.

Several important complexity classes are defined in terms of Boolean circuits, including NC. NC is defined to be the set of Boolean functions that can be decided by uniform Boolean circuits of polynomial size and polylogarithmic depth. Here, the word *uniform* means that there must be some condition on the circuit family so that a description of a circuit can be computed from its index, n.

Formal Definition

In giving a formal definition of Boolean circuits, Vollmer starts by defining a basis set B of Boolean functions, corresponding to the gates allowable in the circuit model. A Boolean circuit over a basis B, with n inputs and m outputs, is then defined as a finite directed acyclic graph. Each vertex corresponds to either a basis function or one of the inputs, and there are a set of exactly m nodes which are labeled as the outputs. The edges must also have some ordering, to distinguish between different arguments to the same Boolean function.

Evaluation of a Circuit

The problem *CIRCUIT-EVAL* takes as input the description of a Boolean circuit and a truth assignment to the variables of the circuit, and accepts only if the circuit evaluate to true. *CIRCUIT-EVAL* is P-complete.

Chapter 7

Time Hierarchy Theorem

In computational complexity theory, the time hierarchy theorems are important statements about time-bounded computation on Turing machines. Informally, these theorems say that given more time, a Turing machine can solve more problems. For example, there are problems that can be solved withn^2 time but not n time.

The time hierarchy theorem for deterministic multi-tape Turing machines was first proven by Richard Stearns and Juris Hartmanis in 1965. It was improved a year later when F. C. Hennie and Richard Stearns improved the efficiency of the Universal Turing machine. As a consequence, for every deterministic time-bounded complexity class, there is a strictly larger time-bounded complexity class, and so the time-bounded hierarchy of complexity classes does not completely collapse. More precisely, the time hierarchy theorem for deterministic Turing machines states that for all time-constructible functions $f(n)$,

$$\mathrm{DTIME}\left(o\left(\frac{f(n)}{\log f(n)}\right)\right) \subsetneq \mathrm{DTIME}(f(n)).$$

The time hierarchy theorem for nondeterministic Turing machines was originally proven by Stephen Cook in 1972. It was improved to its current form via a complex proof by Joel Seiferas, Michael Fischer, and Albert Meyer in 1978. Finally in 1983, Stanislav Žαk achieved the same result with the simple proof taught today. The time hierarchy theorem for nondeterministic Turing machines states that if $g(n)$ is a time-constructible function, and $f(n+1) = o(g(n))$, then

$$\mathrm{NTIME}(f(n)) \subsetneq \mathrm{NTIME}(g(n))$$

The analogous theorems for space are the space hierarchy theorems. A similar theorem is not known for time-bounded probabilistic complexity classes, unless the class also has advice.

Background

Both theorems use the notion of a time-constructible function. A function $f:\mathbb{N}\to\mathbb{N}$ is time-constructible if there exists a deterministic Turing machine such that for every $n\in\mathbb{N}$, if the machine is started with an input of n ones, it will halt after precisely $f(n)$ steps. All polynomials with non-negative integral coefficients are time-constructible, as are exponential functions such as 2^n.

Proof Overview

We need to prove that some time class TIME($g(n)$) is strictly larger than some time class TIME($f(n)$). We do this by constructing a machine which cannot be in TIME($f(n)$), by diagonalization. We then show that the machine is in TIME($g(n)$), using a simulator machine.

Deterministic Time Hierarchy Theorem

Statement

The theorem states that: If $f(n)$ is a time-constructible function, then there exists a decision problemwhich cannot be solved in worst-case deterministic time $f(n)$ but can be solved in worst-case deterministic time $f(n)^2$. In other words, the complexity class DTIME($f(n)$) is a strict subset of DTIME($f(n)^2$). Note that $f(n)$ is at least n, since smaller functions are never time-constructible.

Even more generally, it can be shown that if $f(n)$ is time-constructible, then $\mathrm{DTIME}\left(o\left(\frac{f(n)}{\log f(n)}\right)\right)$ is properly contained in $\mathrm{DTIME}(f(n))$. For example, there are problems solvable in time n^2 but not time n, since n is in $o\left(\frac{n^2}{\log n^2}\right)$.

Proof

We include here a proof that DTIME($f(n)$) is a strict subset of DTIME($f(2n + 1)^3$) as it is simpler. See the bottom of this section for information on how to extend the proof to $f(n)^2$.

To prove this, we first define a language as follows:

$$H_f = \{([M],x) \mid M \text{ accepts } x \text{ in } f(|x|) \text{ steps}\}.$$

Here, M is a deterministic Turing machine, and x is its input (the initial contents of its tape). $[M]$ denotes an input that encodes the Turing machine M. Let m be the size of the tuple ($[M]$, x).

We know that we can decide membership of H_f by way of a deterministic Turing machine that first calculates $f(|x|)$, then writes out a row of 0s of that length, and then uses this row of 0s as a "clock" or "counter" to simulate M for at most that many steps.

At each step, the simulating machine needs to look through the definition of M to decide what the next action would be. It is safe to say that this takes at most $f(m)^3$ operations, so

$$H_f \in \mathsf{TIME}(f(m)^3).$$

The rest of the proof will show that

$$H_f \notin \mathsf{TIME}(f(\lfloor m/2 \rfloor))$$

so that if we substitute $2n + 1$ for m, we get the desired result. Let us assume that H_f is in this time complexity class, and we will attempt to reach a contradiction.

If H_f is in this time complexity class, it means we can construct some machine K which, given some machine description $[M]$ and input x, decides whether the tuple $([M], x)$ is in H_f within $\mathsf{TIME}(f(\lfloor m/2 \rfloor))$.

Therefore we can use this K to construct another machine, N, which takes a machine description $[M]$ and runs K on the tuple $([M], [M])$, and then accepts only if K rejects, and rejects if K accepts. If now $n$is the length of the input to N, then m (the length of the input to K) is twice n plus some delimiter symbol, so $m = 2n + 1$. N's running time is thus $\mathsf{TIME}(f(\lfloor m/2 \rfloor)) = \mathsf{TIME}(f(\lfloor (2n+1)/2 \rfloor)) = \mathsf{TIME}(f(n))$

Now if we feed $[N]$ as input into N itself (which makes n the length of $[N]$) and ask the question whether N accepts its own description as input, we get:

- If N accepts $[N]$ (which we know it does in at most $f(n)$ operations), this means that K rejects $([N], [N])$, so $([N], [N])$ is not in H_f, and thus N does not accept $[N]$ in $f(n)$ steps. Contradiction!
- If N rejects $[N]$ (which we know it does in at most $f(n)$ operations), this means that K accepts $([N], [N])$, so $([N], [N])$ is in H_f, and thus N does accept $[N]$ in $f(n)$ steps. Contradiction!

We thus conclude that the machine K does not exist, and so

$$H_f \notin \mathsf{TIME}(f(\lfloor m/2 \rfloor)).$$

Extension

The reader may have realised that the proof is simpler because we have chosen a simple Turing machine simulation for which we can be certain that

$$H_f \in \mathsf{TIME}(f(m)^3).$$

It has been shown that a more efficient model of simulation exists which establishes that

$$H_f \in \mathsf{TIME}(f(m)\log f(m))$$

but since this model of simulation is rather involved, it is not included here.

Non-deterministic Time Hierarchy Theorem

If $g(n)$ is a time-constructible function, and $f(n+1) = o(g(n))$, then there exists a decision problem which cannot be solved in non-deterministic time $f(n)$ but can be solved in non-deterministic time $g(n)$. In other words, the complexity class NTIME($f(n)$) is a strict subset of NTIME($g(n)$).

Consequences

The time hierarchy theorems guarantee that the deterministic and non-deterministic versions of theexponential hierarchy are genuine hierarchies:

in other words $\mathbf{P} \subset \mathrm{EXPTIME} \subset 2-\mathrm{EXP} \subset$ and

$\mathbf{NP} \subset \mathrm{NEXPTIME} \subset \mathbf{2}-\mathbf{NEXP} \subset$

For example, $\mathbf{P} \subset \mathbf{EXPTIME}\,\mathrm{since}\,\mathbf{P} \subseteq \mathbf{DTIME}(2^n) \subset \mathbf{DTIME}(2^{2n}) \subseteq \mathbf{EXPTIME}$

The theorem also guarantees that there are problems in P requiring arbitrary large exponents to solve; in other words, P does not collapse to DTIME(n^k) for any fixed k. For example, there are problems solvable in n^{5000} time but not n^{4999} time. This is one argument against Cobham's thesis, the convention that P is a practical class of algorithms. If such a collapse did occur, we could deduce thatP $\neq$ PSPACE, since it is a well-known theorem that DTIME($f(n)$) is strictly contained in DSPACE($f(n)$).

However, the time hierarchy theorems provide no means to relate deterministic and non-deterministic complexity, or time and space complexity, so they cast no light on the great unsolved questions ofcomputational complexity theory: whether P and NP, NP and PSPACE, PSPACE and EXPTIME, orEXPTIME and NEXPTIME are equal or not

Space Hierarchy Theorem

In computational complexity theory, the space hierarchy theorems are separation results that show that both deterministic and nondeterministic machines can solve more problems in (asymptotically) more space, subject to certain conditions. For example, a deterministic Turing machine can solve more decision problems in space $n \log n$ than in space n. The somewhat weaker analogous theorems for time are the time hierarchy theorems.

The foundation for the hierarchy theorems lies in the intuition that with either more time or more space comes the ability to compute more functions (or decide more languages). The hierarchy theorems are used to demonstrate that the time and space complexity classes form a hierarchy where classes with tighter bounds contain fewer languages than those with more relaxed bounds. Here we define and prove the space hierarchy theorem.

The space hierarchy theorems rely on the concept of space-constructible functions. The deterministic and nondeterministic space hierarchy theorems state that for all space-constructible functions $f(n)$,

$$\mathrm{SPACE}\big(o(f(n))\big) \subsetneq \mathrm{SPACE}(f(n))$$

where SPACE stands for either DSPACE or NSPACE.

Statement

Formally, a function $f : \mathbb{N} \to \mathbb{N}$ is space-constructible if $f(n) \geq \log n$ and there exists a Turing machine which computes the function $f(n)$ in space $O(f(n))$ when starting with an input 1^n, where 1^n represents a string of n 1s. Most of the common functions that we work with are space-constructible, including polynomials, exponents, and logarithms.

For every space-constructible function $f : \mathbb{N} \to \mathbb{N}$, there exists a language L that is decidable in space $O(f(n))$ but not in space $o(f(n))$.

Proof

The goal here is to define a language that can be decided in space $O(f(n))$ but not space $o(f(n))$. Here we define the language L:

$$L = \{ (\langle M \rangle, 10^k) : M \text{ does not accept}$$
$$(\langle M \rangle, 10^k) \text{ using space } \leq f(|\langle M \rangle, 10^k|) \}$$

Now, for any machine M that decides a language in space $o(f(n))$, L will differ in at least one spot from the language of M, namely at the value of $(\langle M\rangle,10^k)$. The algorithm for deciding the language L is as follows:

1. On an input x, compute $f(\mid x \mid)$ using space-constructibility, and mark off $f(\mid x \mid)$ cells of tape. Whenever an attempt is made to use more than $f(\mid x \mid)$ cells, *reject*.
2. If x is not of the form $\langle M\rangle,10^k$ for some TM M, *reject*.
3. Simulate M on input x for at most $2^{f(\mid x \mid)}$ steps (using $f(\mid x \mid)$ space). If the simulation tries to use more than $f(\mid x \mid)$ space or more than $2^{f(\mid x \mid)}$ operations, then *reject*.
4. If M accepted x during this simulation, then *reject*; otherwise, *accept*.

Note on step 3: Execution is limited to $2^{f(\mid x \mid)}$ steps in order to avoid the case where M does not halt on the input x. That is, the case where M consumes space of only $O(f(x))$ as required, but runs for infinite time.

Comparison and Improvements

The space hierarchy theorem is stronger than the analogous time hierarchy theorems in several ways:

- It only requires s(n) to be at least log n instead of at least n.
- It can separate classes with any asymptotic difference, whereas the time hierarchy theorem requires them to be separated by a logarithmic factor.
- It only requires the function to be space-constructible, not time-constructible.

It seems to be easier to separate classes in space than in time. Indeed, whereas the time hierarchy theorem has seen little remarkable improvement since its inception, the nondeterministic space hierarchy theorem has seen at least one important improvement by Viliam Geffert in his 2003 paper "Space hierarchy theorem revised". This paper made several striking generalizations of the theorem:

- It relaxes the space-constructibility requirement. Instead of merely separating the union classesDSPACE(O(s(n)) and DSPACE(o(s(n)), it separates DSPACE(f(n)) from DSPACE(g(n)) where f(n) is an arbitrary O(s(n)) function and g(n) is a computable o(s(n)) function. These functions need not be space-constructible or even monotone increasing.

- It identifies a unary language, or tally language, which is in one class but not the other. In the original theorem, the separating language was arbitrary.
- It does not require s(n) to be at least log n; it can be any nondeterministically fully space-constructible function.

Corollaries

Corollary 1

For any two functions f_1, $f_2 : \mathbb{N} \rightarrow \mathbb{N}$ where $f_1(n)$ is $o(f_2(n))$ and f_2 is space-constructible, $\text{SPACE}(f_1(n)) \subset \text{SPACE}(f_2(n))$..

This corollary lets us separate various space complexity classes. For any function n^k is space-constructible for any natural number k. Therefore for any two natural numbers $k_1 < k_2$ we can prove $\text{SPACE}(n^{k_1}) \subset \text{SPACE}(n^{k_2})$. We can extend this idea for real numbers in the following corollary. This demonstrates the detailed hierarchy within the PSPACE class.

Corollary 2

For any two real numbers $0 \leq a_1 < a_2$, $\text{SPACE}(n^{a_1}) \subset \text{SPACE}(n^{a_2})$.

Corollary 3

NL $\subset$ *PSPACE.*

Proof

Savitch's theorem shows that NL $\subseteq$ SPACE(log $^2 n$), while the space hierarchy theorem shows that SPACE $\log^2 n) \subsetneq$ (SPACE(n). Thus we get this corollary along with the fact that TQBF NL since TQBF is PSPACE-complete.

This could also be proven using the non-deterministic space hierarchy theorem to show that NL NPSPACE, and using Savitch's theorem to show that PSPACE = NPSPACE.

Corollary 4

PSPACE $\subset$ EXPSPACE.

This last corollary shows the existence of decidable problems that are intractable. In other words their decision procedures must use more than polynomial space.

Corollary 5

There are problems in PSPACE requiring an arbitrarily large

exponent to solve; therefore PSPACEdoes not collapse to DSPACE(n^k) for some constant k.

For the complexity classes defined in this way, it is desirable to prove that relaxing the requirements on (say) computation time indeed defines a bigger set of problems.

In particular, although DTIME(n) is contained in DTIME(n^2), it would be interesting to know if the inclusion is strict. For time and space requirements, the answer to such questions is given by the time and space hierarchy theorems respectively.

They are called hierarchy theorems because they induce a proper hierarchy on the classes defined by constraining the respective resources.

Thus there are pairs of complexity classes such that one is properly included in the other. Having deduced such proper set inclusions, we can proceed to make quantitative statements about how much more additional time or space is needed in order to increase the number of problems that can be solved.

More precisely, the time hierarchy theorem states that

$$\mathrm{DTIME}\big(f(n)\big) \subsetneq \mathrm{DTIME}\big(f(n) \cdot \log^2(f(n))\big).$$

The space hierarchy theorem states that

$$\mathrm{DSPACE}\big(f(n)\big) \subsetneq \mathrm{DSPACE}\big(f(n) \cdot \log(f(n))\big)$$

The time and space hierarchy theorems form the basis for most separation results of complexity classes. For instance, the time hierarchy theorem tells us that P is strictly contained in EXPTIME, and the space hierarchy theorem tells us that L is strictly contained in PSPACE.

Reduction (Complexity

In computability theory and computational complexity theory, a reduction is a transformation of one problem into another problem. Depending on the transformation used this can be used to define complexity classes on a set of problems.

Intuitively, problem A is reducible to problem B if solutions to B exist and give solutions to A whenever A has solutions. Thus, solving A cannot be harder than solving B. We write A $\leq_m$ B, usually with a subscript on the $\leq$ to indicate the type of reduction being used (m : mapping reduction, p : polynomial reduction).

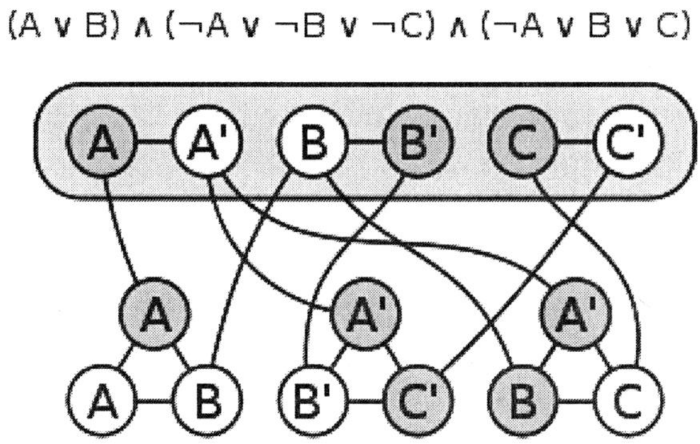

Figure: Example of a reduction from a boolean satisfiability problem to a vertex cover problem. Blue vertices form a vertex cover which corresponds to truth values.

Often we find ourselves trying to solve a problem that is similar to a problem we've already solved. In these cases, often a quick way of solving the new problem is to transform each instance of the new problem into instances of the old problem, solve these using our existing solution, and then use these to obtain our final solution. This is perhaps the most obvious use of reductions.

Another, more subtle use is this: suppose we have a problem that we've proven is hard to solve, and we have a similar new problem. We might suspect that it, too, is hard to solve. We argue by contradiction: suppose the new problem is easy to solve. Then, if we can show that every instance of the old problem can be solved easily by transforming it into instances of the new problem and solving those, we have a contradiction. This establishes that the new problem is also hard.

A very simple example of a reduction is from *multiplication* to *squaring*. Suppose all we know how to do is to add, subtract, take squares, and divide by two. We can use this knowledge, combined with the following formula, to obtain the product of any two numbers:

$$a \times b = \frac{\left((a+b)^2 - a^2 - b^2\right)}{2}$$

We also have a reduction in the other direction; obviously, if we can multiply two numbers, we can square a number. This seems to imply that these two problems are equally hard. This kind of reduction corresponds to Turing reduction.

However, the reduction becomes much harder if we add the restriction that we can only use the squaring function one time, and only at the end. In this case, even if we're allowed to use all the basic arithmetic operations, including multiplication, no reduction exists in general, because we may have to compute an irrational number like $\sqrt{2}$ from rational numbers.

Going in the other direction, however, we can certainly square a number with just one multiplication, only at the end. Using this limited form of reduction, we have shown the unsurprising result that multiplication is harder in general than squaring. This corresponds to many-one reduction.

Definition

Given two subsets A and B of N and a set of functions F from N to N which is closed undercomposition, A is called reducible to B under F if

$$\exists f \in F \,.\, \forall x \in \mathbb{N} \,.\, x \in A \Leftrightarrow f(x) \in B$$

We write

$$A \leq_F B$$

Let S be a subset of P(N) and ≤ a reduction, then S is called closed under ≤ if

$$\forall s \in S \,.\, \forall A \in P(\mathbb{N}) \,.\, A \leq s \Rightarrow A \in S$$

A subset A of N is called hard for S if

$$\forall s \in S \,.\, s \leq A$$

A subset A of N is called complete for S if A is hard for S and A is in S.

Properties

A reduction is a preordering, that is a reflexive and transitive relation, on P(N)×P(N), where P(N) is thepower set of the natural numbers.

Types and Applications of Reductions

As described in the example above, there are two main types of reductions used in computational complexity, the many-one reduction and the Turing reduction. Many-one reductions map *instances* of one problem to *instances* of another; Turing reductions *compute* the solution to one problem, assuming the other problem is easy to solve. A many-one reduction is weaker than a Turing reduction. Weaker reductions are more effective at separating problems, but they have less power, making reductions harder to design.

A problem is complete for a complexity class if every problem in the class reduces to that problem, and it is also in the class itself. In this sense the problem represents the class, since any solution to it can, in combination with the reductions, be used to solve every problem in the class.

However, in order to be useful, reductions must be *easy*. For example, it's quite possible to reduce a difficult-to-solve NP-complete problem like the boolean satisfiability problem to a trivial problem, like determining if a number equals zero, by having the reduction machine solve the problem in exponential time and output zero only if there is a solution. However, this does not achieve much, because even though we can solve the new problem, performing the reduction is just as hard as solving the old problem. Likewise, a reduction computing a noncomputable function can reduce an undecidable problem to a decidable one. As Michael Sipser points out in *Introduction to the Theory of Computation*: "The reduction must be easy, relative to the complexity of typical problems in the class [...] If the reduction itself were difficult to compute, an easy solution to the complete problem wouldn't necessarily yield an easy solution to the problems reducing to it."

Therefore, the appropriate notion of reduction depends on the complexity class being studied. When studying the complexity class NP and harder classes such as the polynomial hierarchy, polynomial-time reductions are used. When studying classes within P such as NC and NL, log-space reductionsare used. Reductions are also used in computability theory to show whether problems are or are not solvable by machines at all; in this case, reductions are restricted only to computable functions.

In case of optimization (maximization or minimization) problems, we often think in terms ofapproximation-preserving reductions. Suppose we have two optimization problems such that instances of one problem can be mapped onto instances of the other, in a way that nearly-optimal solutions to instances of the latter problem can be transformed back to yield nearly-optimal solutions to the former. This way, if we have an optimization algorithm (or approximation algorithm) that finds near-optimal (or optimal) solutions to instances of problem

B, and an efficient approximation-preserving reduction from problem A to problem B, by composition we obtain an optimization algorithm that yields near-optimal solutions to instances of problem A. Approximation-preserving reductions are often used to provehardness of approximation results: if some optimization problem A is hard to approximate (under some complexity assumption) within a factor better than α for some α, and there is a β-approximation-preserving reduction from problem A to problem B, we can conclude that problem B is hard to approximate within factor α/β.

Examples

- To show that a decision problem P is undecidable we must find a reduction from a decision problem which is already known to be undecidable to P. That reduction function must be acomputable function. In particular, we often show that a problem P is undecidable by showing that the halting problem reduces to P.
- The complexity classes P, NP and PSPACE are closed under polynomial-time reductions.
- The complexity classes L, NL, P, NP and PSPACE are closed under log-space reduction.

Detailed Example

The following example shows how to use reduction from the halting problem to prove that a language is undecidable. Suppose $H(M, w)$ is the problem of determining whether a given Turing machine M halts (by accepting or rejecting) on input string w. This language is known to be undecidable. Suppose $E(M)$ is the problem of determining whether the language a given Turing machine M accepts is empty (in other words, whether M accepts any strings at all). We show that E is undecidable by a reduction from$H$.

To obtain a contradiction, suppose R is a decider for E. We will use this to produce a decider S for H(which we know does not exist). Given input M and w (a Turing machine and some input string), define$S(M, w)$ with the following behavior: S creates a Turing machine N that accepts only if the input string to N is w and M halts on input w, and does not halt otherwise. The decider S can now evaluate $R(N)$ to check whether the language accepted by N is empty. If R accepts N, then the language accepted by$N$ is empty, so in particular M does not halt on input w, so S can reject. If R rejects N, then the language accepted by N is nonempty, so M does halt on input w, so S can accept. Thus, if we had a decider R for E, we would be able to produce a decider S for the halting problem $H(M, w)$ for any machine M and input w. Since we know that such an S cannot exist, it follows that the language E is also undecidable.

Polynomial-Time Reduction

In computational complexity theory a polynomial-time reduction is a reduction which is computable by a deterministic Turing machine in polynomial time. If it is a many-one reduction, it is called apolynomial-time many-one reduction, polynomial transformation, or Karp reduction. If it is aTuring reduction, it is called a polynomial-time Turing reduction or Cook reduction.

Polynomial-time reductions are important and widely used because they are powerful enough to perform many transformations between important problems, but still weak enough that polynomial-time reductions from problems in NP or co-NP to problems in P are considered unlikely to exist. This notion of reducibility is used in the standard definitions of several complete complexity classes, such as NP-complete, PSPACE-complete and EXPTIME-complete.

Within the class P, however, polynomial-time reductions are inappropriate, because any problem in P can be polynomial-time reduced (both many-one and Turing) to almost any other problem in P. Thus, for classes within P such as L, NL, NC, and P itself, log-space reductions are used instead.

If a problem has a Karp reduction to a problem in NP, this shows that the problem is in NP. Cook reductions seem to be more powerful than Karp reductions; for example, any problem in co-NP has a Cook reduction to any NP-complete problem, whereas any problems that are in co-NP - NP (assuming they exist) will not have Karp reductions to any problem in NP. While this power is useful for designing reductions, the downside is that certain classes such as NP are not known to be closed under Cook reductions (and are widely believed not to be), so they are not useful for proving that a problem is in NP. However, they are useful for showing that problems are in P and other classes that are closed under such reductions.

Log-space Reduction

In computational complexity theory, a log-space reduction is a reduction computable by adeterministic Turing machine using logarithmic space. Conceptually, this means it can keep a constant number of pointers into the input, along with a logarithmic number of fixed-size integers. Since such a machine has polynomially-many configurations, log-space reductions are alsopolynomial-time reductions.

Log-space reductions are probably weaker than polynomial-time reductions; while any non-empty, non-full language in P is polynomial-time reducible to any other non-empty, non-full language in P, a log-space reduction between a language in NL and a language in L, both subsets of P, would imply the unlikely L = NL. It is an open question if the NP-complete problems are different with respect to log-space and polynomial-time reductions.

Log-space reductions are normally used on languages in P, in which case it usually does not matter whether many-one reductions or Turing reductions are used, since it has been verified that L, SL, NL, and P are all closed under Turing reductions, meaning that Turing reductions can be used to show a problem is in any of these classes. However, other subclasses of P such as NC may not be closed under Turing reductions, and so many-one reductions must be used.

Just as polynomial-time reductions are useless within P and its subclasses, log-space reductions are useless to distinguish problems in L and its subclasses; in particular, almost every problem in L is trivially L-complete under log-space reductions. While even weaker reductions exist, they are not often used in practice, because complexity classes smaller than L (that is, strictly contained or thought to be strictly contained in L) receive relatively little attention.

The tools available to designers of log-space reductions have been greatly expanded by the result that L = SL; see SL for a list of some SL-complete problems that can now be used as subroutines in log-space reductions.

Complete (Complexily)

In computational complexity theory, a computational problem is complete for a complexity class if it is, in a formal sense, one of the "hardest" or "most expressive" problems in the complexity class. If a problem has the property that (if you know how to solve it) it allows you to quickly solve any problem in a complexity class, it is called hard for that class. More formally, a problem p is called hard for a complexity class C under a given type of reduction, if there is a reduction of this type from any problem in C to p. If a problem is both hard for a class, and a member of the class, it is complete for that class (under the given type of reduction).

A problem that is complete for a class C is said to be C-complete, and the class of all problems complete for C is denoted C-complete. The first complete class to be defined and the most well-known is NP-complete, a class that contains many difficult-to-solve problems that arise in practice. Similarly, a problem hard for a class C is called C-hard, e.g. NP-hard.

Normally it is assumed that the reduction in question does not have higher computational complexity than the class itself. Therefore it may be said that if a *C-complete* problem has a "computationally easy" solution, then all problems in "C" have an "easy" solution.

Generally, complexity classes that have a recursive enumeration have known complete problems, whereas those that do not, don't have any known complete problems. For example, NP, co-NP, PLS, PPA all have known natural complete problems, while RP, ZPP, BPP and TFNP do not have any known complete problems (although such a problem may be discovered in the future).

There are classes without complete problems. For example, Sipser showed that there is a language Msuch that BPP^M (BPP with oracle M) has no complete problems.

NP-complete

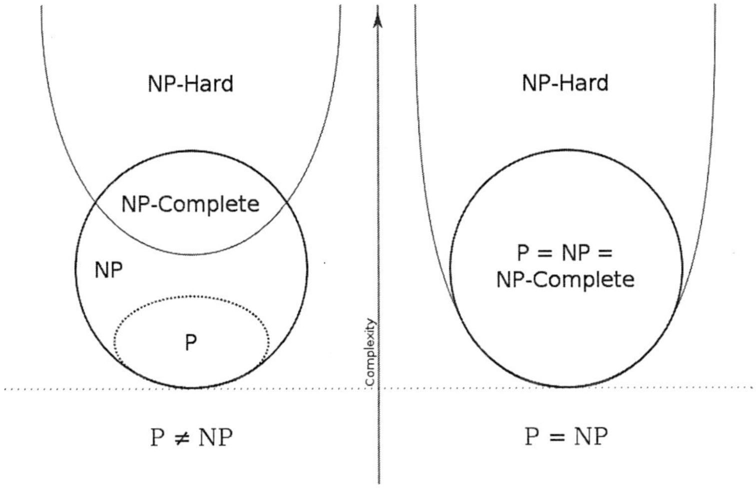

Figure: Euler diagram for P, NP, NP-complete, and NP-hard set of problems

In computational complexity theory, thecomplexity class NP-complete (abbreviatedNP-C or NPC) is a class of decision problems. A decision problem L is NP-complete if it is in the set of NP problems so that any given solution to the decision problem can be verified in polynomial time, and also in the set of NP-hard problems so that any NP problem can be converted into $L$by a transformation of the inputs in polynomial time.

Although any given solution to such a problem can be verified quickly, there is no known efficient way to locate a solution in the first place; indeed, the most notable characteristic of NP-complete problems is that no fast solution to them is known. That is, the time required to solve the problem using any currently known algorithmincreases very quickly as the size of the problem grows. As a result, the time required to solve even moderately large versions of many of these problems easily reaches into the billions or trillions of years, using any amount of computing power available today. As a consequence, determining whether or not it is possible to solve these problems quickly, called the P versus NP problem, is one of the principal unsolved problems in computer science today.

While a method for computing the solutions to NP-complete problems using a reasonable amount of time remains undiscovered, computer scientists and programmers still frequently encounter NP-complete problems. NP-complete problems are often addressed by using approximation algorithms.

Formal Overview

NP-complete is a subset of NP, the set of all decision problems whose solutions can be verified in polynomial time; *NP* may be equivalently defined as the set of decision problems that can be solved in polynomial time on a nondeterministic Turing machine. A problem p in NP is also in NPC if and only ifevery other problem in NP can be transformed into p in polynomial time. NP-complete can also be used as an adjective: problems in the class NP-complete are known as NP-complete problems.

NP-complete problems are studied because the ability to quickly verify solutions to a problem (NP) seems to correlate with the ability to quickly solve that problem (P). It is not known whether every problem in NP can be quickly solved—this is called the P = NP problem. But if *any single problem* in NP-complete can be solved quickly, then *every problem in NP* can also be quickly solved, because the definition of an NP-complete problem states that every problem in NP must be quickly reducible to every problem in NP-complete (that is, it can be reduced in polynomial time). Because of this, it is often said that the NP-complete problems are *harder* or *more difficult* than NP problems in general.

Formal Definition of NP-completeness

A decision problem C is NP-complete if:

1. is in NP, and
2. Every problem in NP is reducible to in polynomial time.

can be shown to be in NP by demonstrating that a candidate solution to can be verified in polynomial time.

Note that a problem satisfying condition 2 is said to be NP-hard, whether or not it satisfies condition 1.

A consequence of this definition is that if we had a polynomial time algorithm (on a UTM, or any otherTuring-equivalent abstract machine) for , we could solve all problems in NP in polynomial time.

Background

The concept of *NP-complete* was introduced in 1971 by Stephen Cook in a paper entitled *The complexity of theorem-proving procedures* on pages 151–158 of the *Proceedings of the 3rd Annual ACM Symposium on Theory of Computing*, though the term *NP-complete* did not appear anywhere in his paper. At that computer science conference, there was a fierce debate among the computer scientists about whether NP-complete problems could be solved in polynomial time on a deterministicTuring machine. John Hopcroft brought everyone at the conference to a consensus that the question of whether NP-complete problems are solvable in polynomial time should be put off to be solved at some later date, since nobody had any formal proofs for their claims one way or the other. This is known as the question of whether P=NP.

Nobody has yet been able to determine conclusively whether NP-complete problems are in fact solvable in polynomial time, making this one of the great unsolved problems of mathematics. The Clay Mathematics Institute is offering a US$1 million reward to anyone who has a formal proof that P=NP or that P≠NP.

In the celebrated Cook-Levin theorem (independently proved by Leonid Levin), Cook proved that theBoolean satisfiability problem is NP-complete (a simpler, but still highly technical proof of this is available). In 1972, Richard Karp proved that several other problems were also NP-complete (seeKarp's 21 NP-complete problems); thus there is a class of NP-complete problems (besides the Boolean satisfiability problem). Since Cook's original results, thousands of other problems have been shown to be NP-complete by reductions from other problems previously shown to be NP-complete; many of these problems are collected in Garey and Johnson's 1979 book *Computers and Intractability: A Guide to the Theory of NP-Completeness*.

NP-complete Problems

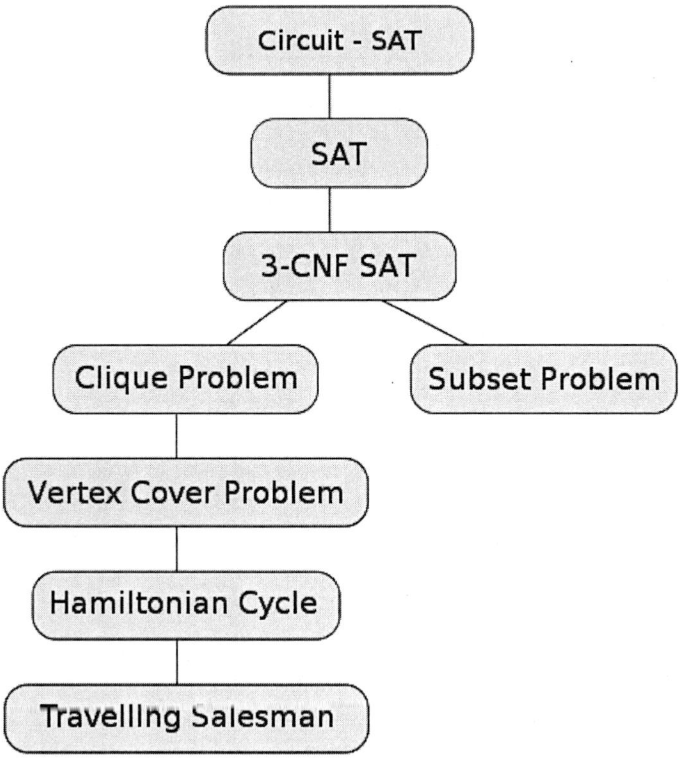

Figure: Some NP-complete problems, indicating the reductions typically used to prove their NP-completeness

An interesting example is the graph isomorphism problem, the graph theory problem of determining whether a graph isomorphism exists between two graphs. Two graphs areisomorphic if one can be transformed into the other simply by renaming vertices. Consider these two problems:

- Graph Isomorphism: Is graph G_1 isomorphic to graph G_2?
- Subgraph Isomorphism: Is graph G_1 isomorphic to a subgraph of graph G_2?

The Subgraph Isomorphism problem is NP-complete. The graph isomorphism problem is suspected to be neither in P nor NP-complete, though it is in NP. This is an example of a problem that is thought to be hard, but isn't thought to be NP-complete.

The easiest way to prove that some new problem is NP-complete is first to prove that it is in NP, and then to reduce some known NP-complete problem to it. Therefore, it is useful to know a variety of NP-complete problems. The list below contains some well-known problems that are NP-complete when expressed as decision problems.

- Boolean satisfiability problem (Sat.)
- N-puzzle
- Knapsack problem
- Hamiltonian path problem
- Travelling salesman problem
- Subgraph isomorphism problem
- Subset sum problem
- Clique problem
- Vertex cover problem
- Independent set problem
- Dominating set problem
- Graph coloring problem.

To the right is a diagram of some of the problems and the reductions typically used to prove their NP-completeness. In this diagram, an arrow from one problem to another indicates the direction of the reduction. Note that this diagram is misleading as a description of the mathematical relationship between these problems, as there exists a polynomial-time reduction between any two NP-complete problems; but it indicates where demonstrating this polynomial-time reduction has been easiest.

There is often only a small difference between a problem in P and an NP-complete problem. For example, the 3-satisfiability problem, a restriction of the boolean satisfiability problem, remains NP-complete, whereas the slightly more restricted 2-satisfiability problem is in P (specifically, NL-complete), and the slightly more general max. 2-sat. problem is again NP-complete.

Determining whether a graph can be colored with 2 colors is in P, but with 3 colors is NP-complete, even when restricted to planar graphs. Determining if a graph is a cycle or is bipartite is very easy (in L), but finding a maximum bipartite or a maximum cycle subgraph is NP-complete.

A solution of theknapsack problem within any fixed percentage of the optimal solution can be computed in polynomial time, but finding the optimal solution is NP-complete.

Solving NP-complete Problems

At present, all known algorithms for NP-complete problems require time that is superpolynomial in the input size, and it is unknown whether there are any faster algorithms.

The following techniques can be applied to solve computational problems in general, and they often give rise to substantially faster algorithms:

- Approximation: Instead of searching for an optimal solution, search for an "almost" optimal one.
- Randomization: Use randomness to get a faster average running time, and allow the algorithm to fail with some small probability. See Monte Carlo method.
- Restriction: By restricting the structure of the input (e.g., to planar graphs), faster algorithms are usually possible.
- Parameterization: Often there are fast algorithms if certain parameters of the input are fixed.
- Heuristic: An algorithm that works "reasonably well" in many cases, but for which there is no proof that it is both always fast and always produces a good result. Metaheuristic approaches are often used.

One example of a heuristic algorithm is a suboptimal $O(n \log n)$ greedy coloring algorithm used forgraph coloring during the register allocation phase of some compilers, a technique called graph-coloring global register allocation. Each vertex is a variable, edges are drawn between variables which are being used at the same time, and colors indicate the register assigned to each variable. Because most RISC machines have a fairly large number of general-purpose registers, even a heuristic approach is effective for this application.

Completeness Under Different Types of Reduction

In the definition of NP-complete given above, the term *reduction* was used in the technical meaning of a polynomial-time many-one reduction.

Another type of reduction is polynomial-time Turing reduction. A problem X is polynomial-time Turing-reducible to a problem Y if, given a subroutine that solves in polynomial time, one could write a program that calls this subroutine and solves in polynomial time. This contrasts with many-one reducibility, which has the restriction that the program can only call the subroutine once, and the return value of the subroutine must be the return value of the program.

If one defines the analogue to NP-complete with Turing reductions instead of many-one reductions, the resulting set of problems won't be smaller than NP-complete; it is an open question whether it will be any larger.

Another type of reduction that is also often used to define NP-completeness is the logarithmic-space many-one reduction which is a many-one reduction that can be computed with only a logarithmic amount of space. Since every computation that can be done in logarithmic space can also be done in polynomial time it follows that if there is a logarithmic-space many-one reduction then there is also a polynomial-time many-one reduction. This type of reduction is more refined than the more usual polynomial-time many-one reductions and it allows us to distinguish more classes such as P-complete. Whether under these types of reductions the definition of NP-complete changes is still an open problem. All currently known NP-complete problems are NP-complete under log space reductions. Indeed, all currently known NP-complete problems remain NP-complete even under much weaker reductions. It is known, however, that AC^0 reductions define a strictly smaller class than polynomial-time reductions.

Naming

According to Don Knuth, the name "NP-complete" was popularized by Alfred Aho, John Hopcroft andJeffrey Ullman in their celebrated textbook "The Design and Analysis of Computer Algorithms". He reports that they introduced the change in the galley proofs for the book (from "polynomially-complete"), in accordance with the results of a poll he had conducted of the Theoretical Computer Science community. Other suggestions made in the poll included "Herculean", "formidable",Steiglitz's "hard-boiled" in honor of Cook, and Shen Lin's acronym "PET", which stood for "probably exponential time", but depending on which way the P versus NP problem went, could stand for "provably exponential time" or "previously exponential time".

Common Misconceptions

The following misconceptions are frequent.

- *"NP-complete problems are the most difficult known problems."* Since NP-complete problems are in NP, their running time is at most exponential. However, some problems provably require more time, for example Presburger arithmetic.
- *"NP-complete problems are difficult because there are so many different solutions."* On the one hand, there are many problems that have a solution space just as large, but can be solved in polynomial time (for example minimum spanning tree). On the other hand, there are NP-problems with at most one solution that are NP-hard under randomized polynomial-time reduction (seeValiant–Vazirani theorem).
- *"Solving NP-complete problems requires exponential time."* First, this would imply P `" NP, which is still an unsolved question. Further, some NP-complete problems actually have algorithms running in superpolynomial, but subexponential time. For example, the Independent set andDominating set problems are NP-complete when restricted to planar graphs, but can be solved in subexponential time on planar graphs using the planar separator theorem.
- *"All instances of an NP-complete problem are difficult."* Often some instances, or even almost all instances, may be easy to solve within polynomial time.

Many complexity classes are defined using the concept of a reduction. A reduction is a transformation of one problem into another problem. It captures the informal notion of a problem being at least as difficult as another problem. For instance, if a problem X can be solved using an algorithm for Y, X is no more difficult than Y, and we say that X *reduces* to Y. There are many different types of reductions, based on the method of reduction, such as Cook reductions, Karp reductions and Levin reductions, and the bound on the complexity of reductions, such as polynomial-time reductions or log-space reductions.

The most commonly used reduction is a polynomial-time reduction. This means that the reduction process takes polynomial time. For example, the problem of squaring an integer can be reduced to the problem of multiplying two integers. This means an algorithm for multiplying two integers can be used to square an integer. Indeed, this can be done by giving the same input to both inputs of the multiplication algorithm. Thus we see that squaring is not more difficult than multiplication, since squaring can be reduced to multiplication.

This motivates the concept of a problem being hard for a complexity class. A problem X is *hard* for a class of problems C if every problem in C can be reduced to X. Thus no problem in C is harder than X, since an algorithm for X allows us to solve any problem in C. Of course, the notion of hard problems depends on the type of reduction being used. For complexity classes larger than P, polynomial-time reductions are commonly used. In particular, the set of problems that are hard for NP is the set of NP-hard problems.

If a problem X is in C and hard for C, then X is said to be *complete* for C. This means that X is the hardest problem in C. (Since many problems could be equally hard, one might say that X is one of the hardest problems in C.) Thus the class of NP-complete problems contains the most difficult problems in NP, in the sense that they are the ones most likely not to be in P. Because the problem P = NP is not solved, being able to reduce a known NP-complete problem, Π_2, to another problem, Π_1, would indicate that there is no known polynomial-time solution for Π_1. This is because a polynomial-time solution to Π_1 would yield a polynomial-time solution to Π_2. Similarly, because all NP problems can be reduced to the set, finding an NP-complete problem that can be solved in polynomial time would mean that P = NP.

Important Open Problems

P Versus NP Problem

The P versus NP problem is a major unsolved problem in computer science. Informally, it asks whether every problem whose solution can be efficiently checked by a computer can also be efficiently solved by a computer. It was introduced in 1971 byStephen Cook in his seminal paper "The complexity of theorem proving procedures" and is considered by many to be the most important open problem in the field. It is one of the seven Millennium Prize Problems selected by the Clay Mathematics Institute to carry a US$ 1,000,000 prize for the first correct solution.

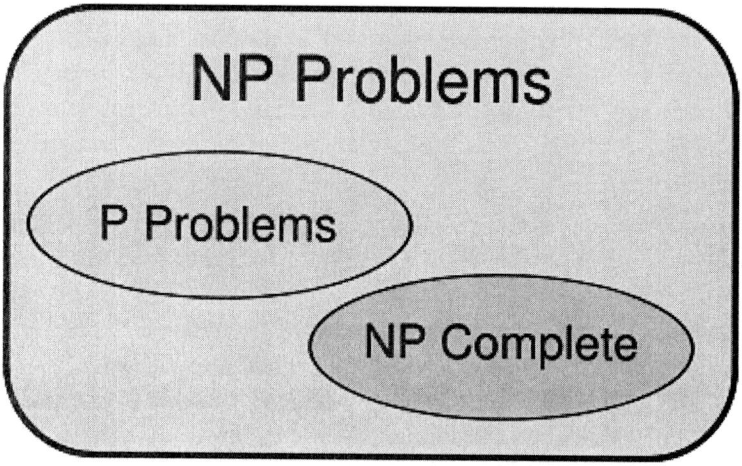

Figure: Diagram of complexity classes provided that P≠ NP. The existence of problems within NP but outside both P and NP-complete, under that assumption, was established by Ladner's theorem.

In essence, the P = NP problem can be restated as the following question:

Suppose that solutions to a problem can be verified quickly. Then, can the solutions themselves also be computed quickly?

The theoretical notion of *quickly* used here is an algorithm that runs in polynomial time. The general class of questions for which some algorithm can provide an answer in polynomial time is called "class P" or just "P". For some questions, there is no known way to find an answer quickly, but if one is provided with information showing what the answer is, it may be possible to verify the answer quickly. The class of questions for which an answer can be verified in polynomial time is called NP.

Consider the subset sum problem, an example of a problem that is easy to verify, but whose answer may be difficult to compute. Given a set of integers, does some nonempty subset of them sum to 0? For instance, does a subset of the set{"2, "3, 15, 14, 7, "10} add up to 0?

The answer "yes, because {"2, "3, "10, 15} add up to zero" can be quickly verified with three additions. However, there is no known algorithm to find such a subset in polynomial time (there is one, however, in exponential time, which consists of 2^n-1 tries), and indeed such an algorithm cannot exist if the two complexity classes are not the same; hence this problem is in NP (quickly checkable) but not necessarily in P (quickly solvable).

An answer to the P = NP question would determine whether problems like the subset-sum problem that can be verified in polynomial time can also be solved in polynomial time. If it turned out that Pdoes not equal NP, it would mean that there are problems in NP (such as NP-complete problems) that are harder to compute than to verify: they could not be solved in polynomial time, but the answer could be verified in polynomial time.

Context

The relation between the complexity classes P and NP is studied in computational complexity theory, the part of the theory of computation dealing with the resources required during computation to solve a given problem. The most common resources are time (how many steps it takes to solve a problem) and space (how much memory it takes to solve a problem).

In such analysis, a model of the computer for which time must be analyzed is required. Typically such models assume that the computer is *deterministic* (given the computer's present state and any inputs, there is only one possible action that the computer might take) and *sequential* (it performs actions one after the other).

In this theory, the class P consists of all those *decision problems* (defined below) that can be solved on a deterministic sequential machine in an amount of time that is polynomial in the size of the input; the class NP consists of all those decision problems whose positive solutions can be verified inpolynomial time given the right information, or equivalently, whose solution can be found in polynomial time on a non-deterministic machine. Clearly, P †" NP. Arguably the biggest open question intheoretical computer science concerns the relationship between those two classes:

Is P equal to NP?

In a 2002 poll of 100 researchers, 61 believed the answer to be no, 9 believed the answer is yes, and 22 were unsure; 8 believed the question may be independent of the currently accepted axioms and so impossible to prove or disprove.

NP-complete

To attack the P = NP question the concept ofNP-completeness is very useful. NP-complete problems are a set of problems to which any other NP-problem can be reduced in polynomial time, but whose solution may still be verified in polynomial time.

Informally, an NP-complete problem is at least as "tough" as any other problem in NP. NP-hardproblems are those at least as hard as NP-complete problems, i.e., all NP-problems can be reduced (in polynomial time) to them. NP-hard problems need not be in NP, i.e., they need not have solutions verifiable in polynomial time.

For instance, the decision problem version of the travelling salesman problem is NP-complete, so *any*instance of *any* problem in NP can be transformed mechanically into an instance of the traveling salesman problem, in polynomial time. The traveling salesman problem is one of many such NP-complete problems. If any NP-complete problem is in P, then it would follow that P = NP. Unfortunately, many important problems have been shown to be NP-complete, and as of 2011 not a single fast algorithm for any of them is known.

Based on the definition alone it is not obvious that NP-complete problems exist. A trivial and contrivedNP-complete problem can be formulated as: given a description of a Turing machine M guaranteed to halt in polynomial time, does there exist a polynomial-size input that M will accept? It is in NPbecause (given an input) it is simple to check whether or not M accepts the input by simulating M; it isNP-complete because the verifier for any particular instance of a problem in NP can be encoded as a polynomial-time machine M that takes the solution to be verified as input. Then the question of whether the instance is a yes or no instance is determined by whether a valid input exists.

The first natural problem proven to be NP-complete was the Boolean satisfiability problem. This result came to be known as Cook–Levin theorem; its proof that satisfiability is NP-complete contains technical details about Turing machines as they relate to the definition of NP. However, after this problem was proved to be NP-complete, proof by reduction provided a simpler way to show that many other problems are in this class. Thus, a vast class of seemingly unrelated problems are all reducible to one another, and are in a sense "the same problem".

Harder Problems

Although it is unknown whether P = NP, problems outside of P are known. A number of succinct problems (problems that operate not on normal input, but on a computational description of the input) are known to be EXPTIME-complete. Because it can be shown that P $\subsetneq$ EXPTIME, these problems are outside P, and so require more than polynomial time. In fact, by the time hierarchy theorem, they cannot be solved in significantly less than exponential time. Examples include finding a perfect strategy for chess (on an N×N board) and some other board games.

The problem of deciding the truth of a statement in Presburger arithmetic requires even more time. Fischer and Rabin proved in 1974 that every algorithm that decides the truth of Presburger statements has a runtime of at least $2^{2^{cn}}$ for some constant c. Here, n is the length of the Presburger statement. Hence, the problem is known to need more than exponential run time. Even more difficult are theundecidable problems, such as the halting problem. They cannot be completely solved by any algorithm, in the sense that for any particular algorithm there is at least one input for which that algorithm will not produce the right answer; it will either produce the wrong answer, finish without giving a conclusive answer, or otherwise run forever without producing any answer at all.

Problems in NP not Known to be in P or NP-complete

It was shown by Ladner that if P ≠ NP then there exist problems in NP that are neither in P nor NP-complete. Such problems are called NP-intermediate problems. The graph isomorphism problem, the discrete logarithm problem and the integer factorization problem are examples of problems believed to be NP-intermediate. They are some of the very few NP problems not known to be in P or to be NP-complete.

The graph isomorphism problem is the computational problem of determining whether two finite graphsare isomorphic. An important unsolved problem in complexity theory is whether the graph isomorphism problem is in P, NP-complete, or NP-intermediate. The answer is not known, but it is believed that the problem is at least not NP-complete. If graph isomorphism is NP-complete, the polynomial time hierarchy collapses to its second level. Since it is widely believed that the polynomial hierarchy does not collapse to any finite level, it is believed that graph isomorphism is not NP-complete. The best algorithm for this problem, due to Laszlo Babai and Eugene Luks has run time $2^{O(\text{“}(n \log n))}$ for graphs with n vertices.

The integer factorization problem is the computational problem of determining the prime factorization of a given integer. Phrased as a decision problem, it is the problem of deciding whether the input has a factor less than k. No efficient integer factorization algorithm is known, and this fact forms the basis of several modern cryptographic systems, such as the RSA algorithm.

The integer factorization problem is in NP and in co-NP (and even in UP and co-UP). If the problem is NP-complete, the polynomial time hierarchy will collapse to its first level (i.e., NP = co-NP). The best known algorithm for integer factorization is the general number field sieve, which takes expected time $O(e^{(64/9)1/3(n.\log 2)1/3(\log (n.\log 2))2/3})$ to factor an n-bit integer.

However, the best known quantum algorithm for this problem, Shor's algorithm, does run in polynomial time. Unfortunately, this fact doesn't say much about where the problem lies with respect to non-quantum complexity classes.

Does P mean "Easy"?

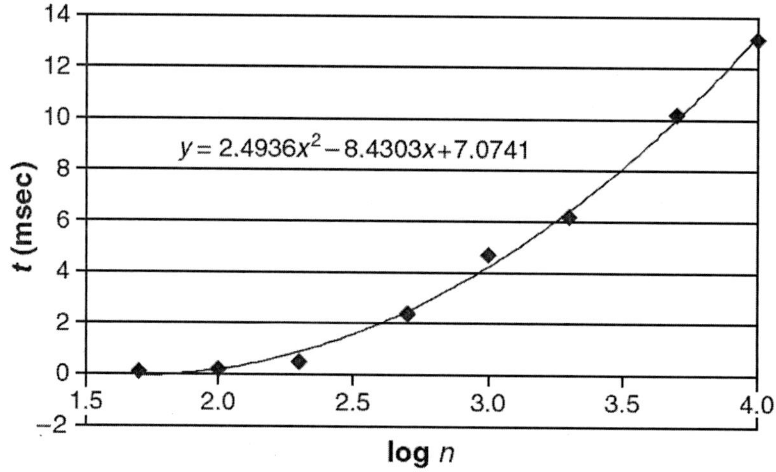

Figure: The graph shows time (average of 100 instances in msec using a 933 MHz Pentium III) vs.problem size for knapsack problems for a state-of-the-art specialized algorithm. Quadratic fit suggests that empirical algorithmic complexity for instances with 50–10,000 variables is $O((\log n)^2)$.

All of the above discussion has assumed that P means "easy" and "not in P" means "hard", an assumption known as *Cobham's thesis.* It is a common and reasonably accurate assumption in complexity theory, however it has some caveats.

First, it is not always true in practice. A theoretical polynomial algorithm may have extremely large constant factors or exponents thus rendering it impractical. On the other hand, even if a problem is shown to be NP-complete, and even if P ≠ NP, there may still be effective approaches to tackling the problem in practice. There are algorithms for many NP-complete problems, such as the knapsack problem, thetraveling salesman problem and the boolean satisfiability problem, that can solve to optimality many real-world instances in reasonable time. The empirical average-case complexity (time vs. problem size) of such algorithms can be surprisingly low. A famous example is the simplex algorithm in linear programming, which works surprisingly well in practice; despite having exponential worst-case time complexity it runs on par with the best known polynomial-time algorithms.

Second, there are types of computations which do not conform to the Turing machine model on whichP and NP are defined, such as quantum computation and randomized algorithms.

Reasons to Believe P ≠ NP

According to a poll, many computer scientists believe that P ≠ NP. A key reason for this belief is that after decades of studying these problems no one has been able to find a polynomial-time algorithm for any of more than 3000 important known NP-complete problems (see List of NP-complete problems). These algorithms were sought long before the concept of NP-completeness was even defined (Karp's 21 NP-complete problems, among the first found, were all well-known existing problems at the time they were shown to be NP-complete). Furthermore, the result P = NP would imply many other startling results that are currently believed to be false, such as NP = co-NP and P =PH.

It is also intuitively argued that the existence of problems that are hard to solve but for which the solutions are easy to verify matches real-world experience.

If P = NP, then the world would be a profoundly different place than we usually assume it to be. There would be no special value in "creative leaps," no fundamental gap between solving a problem and recognizing the solution once it's found. Everyone who could appreciate a symphony would be Mozart; everyone who could follow a step-by-step argument would be Gauss.. — Scott Aaronson, MIT

On the other hand, some researchers believe that there is overconfidence in believing P ≠ NP and that researchers should explore proofs of P = NP as well. For example, in 2002 these statements were made:

The main argument in favor of P ≠ NP is the total lack of fundamental progress in the area of exhaustive search. This is, in my opinion, a very weak argument. The space of algorithms is very large and we are only at the beginning of its exploration. [. . .] The resolution of Fermat's Last Theorem also shows that very simple questions may be settled only by very deep theories. —Moshe Y. Vardi, Rice University

Being attached to a speculation is not a good guide to research planning. One should always try both directions of every problem. Prejudice has caused famous mathematicians to fail to solve famous problems whose solution was opposite to their expectations, even though they had developed all the methods required. —Anil Nerode, Cornell University

Consequences of the Resolution of the Problem

One of the reasons the problem attracts so much attention is the consequences of the answer. Either direction of resolution would advance theory enormously, and perhaps have huge practical consequences as well.

P = NP

A proof that P = NP could have stunning practical consequences, if the proof leads to efficient methods for solving some of the important problems in NP. It is also possible that a proof would not lead directly to efficient methods, perhaps if the proof is non-constructive, or the size of the bounding polynomial is too big to be efficient in practice. The consequences, both positive and negative, arise since various NP-complete problems are fundamental in many fields.

Cryptography, for example, relies on certain problems being difficult. A constructive and efficient solution to an NP-complete problem such as 3-SAT would break most existing cryptosystems including public-key cryptography, a foundation for many modern security applications such as secure economic transactions over the Internet, and symmetric ciphers such as AES or 3DES,used for the encryption of communications data. These would need to be modified or replaced byinformation-theoretically secure solutions.

On the other hand, there are enormous positive consequences that would follow from rendering tractable many currently mathematically intractable problems.

For instance, many problems inoperations research are NP-complete, such as some types of integer programming, and the travelling salesman problem, to name two of the most famous examples. Efficient solutions to these problems would have enormous implications for logistics. Many other important problems, such as some problems in protein structure prediction, are also NP-complete; if these problems were efficiently solvable it could spur considerable advances in biology.

But such changes may pale in significance compared to the revolution an efficient method for solvingNP-complete problems would cause in mathematics itself. According to Stephen Cook,

...it would transform mathematics by allowing a computer to find a formal proof of any theorem which has a proof of a reasonable length, since formal proofs can easily be recognized in polynomial time. Example problems may well include all of the CMI prize problems.

Research mathematicians spend their careers trying to prove theorems, and some proofs have taken decades or even centuries to find after problems have been stated—for instance, Fermat's Last Theorem took over three centuries to prove. A method that is guaranteed to find proofs to theorems, should one exist of a "reasonable" size, would essentially end this struggle.

P ≠ NP

A proof that showed that P ≠ NP would lack the practical computational benefits of a proof that P =NP, but would nevertheless represent a very significant advance in computational complexity theory and provide guidance for future research. It would allow one to show in a formal way that many common problems cannot be solved efficiently, so that the attention of researchers can be focused on partial solutions or solutions to other problems. Due to widespread belief in P ≠ NP, much of this focusing of research has already taken place. Also P ≠ NP still leaves open the average-case complexity of hard problems in NP. For example, it is possible that SAT requires exponential time in the worst case, but that almost all randomly selected instances of it are efficiently solvable. Russell Impagliazzo has described five hypothetical "worlds" that could result from different possible resolutions to the average-case complexity question.

These range from "Algorithmica", where P = NP and problems like SAT can be solved efficiently in all instances, to "Cryptomania", where P ≠ NP and generating hard instances of problems outside P is easy, with three intermediate possibilities reflecting different possible distributions of difficulty over instances of NP-hard problems.

The "world" where P ≠ NP but all problems in NP are tractable in the average case is called "Heuristica" in the paper. A Princeton University workshop in 2009 studied the status of the five worlds.

Results About Difficulty of Proof

Although the P = NP? problem itself remains open, despite a million-dollar prize and a huge amount of dedicated research, efforts to solve the problem have led to several new techniques.

In particular, some of the most fruitful research related to the P = NP problem has been in showing that existing proof techniques are not powerful enough to answer the question, thus suggesting that novel technical approaches are required.

As additional evidence for the difficulty of the problem, essentially all known proof techniques incomputational complexity theory fall into one of the following classifications, each of which is known to be insufficient to prove that P '+ NP:

Classification	*Definition*
Relativizing proofs	Imagine a world wher2€ery algorithm is allowed to make queries to some fixed subroutine called an oracle, and the running time of the oracle is not counted against the running time of the algorithm. Most proofs (especially classical ones) apply uniformly in a world with oracles regardless of what the oracle does. These proofs are called *relativizing*. In 1975, Baker, Gill, and Solovay showed that P = NPwith respect to some oracles, while P ≠ NP for other oracles. Since relativizing proofs can only prove statements that are uniformly true with respect to all possible oracles, this showed that relativizing techniques cannot resolve P = NP.

Natural proofs	In 1993, Alexander Razborov and Steven Rudich defined a general class of proof techniques for circuit complexity lower bounds, called *natural proofs*. At the time all previously known circuit lower bounds were natural, and circuit complexity was considered a very promising approach for resolving P = NP. However, Razborov and Rudich showed that, if one-way functions exist, then no natural proof method can distinguish between P and NP. Although one-way functions have never been formally proven to exist, most mathematicians believe that they do, and a proof or disproof of their existence would be a much stronger statement than the quantification of P relative to NP. Thus it is unlikely that natural proofs alone can resolve P = NP.
Algebrizing proofs	After the Baker-Gill-Solovay result, new non-relativizing proof techniques were successfully used to prove that IP = PSPACE. However, in 2008, Scott Aaronsonand Avi Wigderson showed that the main technical tool used in the IP = PSPACEproof, known as *arithmetization*, was also insufficient to resolve P = NP.

These barriers are another reason why NP-complete problems are useful: if a polynomial-time algorithm can be demonstrated for an NP-complete problem, this would solve the P = NP problem in a way not excluded by the above results.

These barriers have also led some computer scientists to suggest that the P versus NP problem may be independent of standard axiom systems like ZFC (cannot be proved or disproved within them). The interpretation of an independence result could be that either no polynomial-time algorithm exists for any NP-complete problem, and such a proof cannot be constructed in (e.g.) ZFC, or that polynomial-time algorithms for NP-complete problems may exist, but it's impossible to prove in ZFC that such algorithms are correct. However, if it can be shown, using techniques of the sort that are currently known to be applicable, that the problem cannot be decided even with much weaker assumptions extending the Peano axioms (PA) for integer arithmetic, then there would necessarily exist nearly-polynomial-time algorithms for every problem in NP. Therefore, if one believes (as most complexity theorists do) that not all problems in NP have efficient algorithms, it would follow that proofs of independence using those techniques cannot be possible. Additionally, this result implies that proving independence from PA or ZFC using currently known techniques is no easier than proving the existence of efficient algorithms for all problems in NP.

Claimed Solutions

While the P versus NP problem is generally considered unsolved, many amateur and some professional researchers have claimed solutions. Woeginger (2010) has a comprehensive list. An August 2010 claim of proof that P ≠ NP, by Vinay Deolalikar, researcher at HP Labs, Palo Alto, received heavy Internet and press attention after being initially described as "seem[ing] to be a relatively serious attempt" by two leading specialists. The proof has been reviewed publicly by academics, and Neil Immerman, an expert in the field, had pointed out two possibly fatal errors in the proof. As of September 15, 2010, Deolalikar was reported to be working on a detailed expansion of his attempted proof. However, the general consensus amongst theoretical computer scientists is now that the attempted proof is neither correct nor a significant advancement in our understanding of the problem.

Logical Characterizations

The P = NP problem can be restated in terms of expressible certain classes of logical statements, as a result of work in descriptive complexity. All languages (of finite structures with a fixed signatureincluding a linear order relation) in P can be expressed in first-order logic with the addition of a suitable least fixed-point combinator (effectively, this, in combination with the order, allows the definition of recursive functions); indeed, (as long as the signature contains at least one predicate or function in addition to the distinguished order relation [so that the amount of space taken to store such finite structures is actually polynomial in the number of elements in the structure]), this precisely characterizes P. Similarly, NP is the set of languages expressible in existential second-order logic—that is, second-order logic restricted to exclude universal quantification over relations, functions, and subsets. The languages in the polynomial hierarchy, PH, correspond to all of second-order logic. Thus, the question "is P a proper subset of NP" can be reformulated as "is existential second-order logic able to describe languages (of finite linearly ordered structures with nontrivial signature) that first-order logic with least fixed point cannot?". The word "existential" can even be dropped from the previous characterization, since P = NP if and only if P = PH (as the former would establish that NP = co-NP, which in turn implies that NP = PH). PSPACE = NPSPACE as established Savitch's theorem, this follows directly from the fact that the square of a polynomial function is still a polynomial function. However, it is believed, but not proven, that a similar relationship may not exist between the polynomial time complexity classes P and NP, so the question is still open.

Polynomial-Time Algorithms

No algorithm for any NP-complete problem is known to run in polynomial time. However, there are algorithms for NP-complete problems with the property that if P = NP, then the algorithm runs in polynomial time (although with enormous constants, making the algorithm impractical). The following algorithm, due to Levin,□ is such an example. It correctly accepts the NP-complete language SUBSET-SUM, and runs in polynomial time if and only if P = NP:

```
// Algorithm that accepts the NP-complete language SUBSET-SUM.
//
// This is a polynomial-time algorithm if and only if P=NP.
//
// "Polynomial-time" means it returns "yes" in polynomial time when
// the answer should be "yes", and runs forever when it is "no".
//
// Input: S = a finite set of integers
// Output: "yes" if any subset of S adds up to 0.
// Runs forever with no output otherwise.
// Note: "Program number P" is the program obtained by
// writing the integer P in binary, then
// considering that string of bits to be a
// program. Every possible program can be
// generated this way, though most do nothing
// because of syntax errors.
FOR N = 1...infinity
 FOR P = 1...N
```

```
Run program number P for N steps with input S
IF the program outputs a list of distinct integers
  AND the integers are all in S
  AND the integers sum to 0
THEN
  OUTPUT "yes" and HALT
```

If, and only if, P = NP, then this is a polynomial-time algorithm accepting an NP-complete language. "Accepting" means it gives "yes" answers in polynomial time, but is allowed to run forever when the answer is "no".

This algorithm is enormously impractical, even if P = NP. If the shortest program that can solve SUBSET-SUM in polynomial time is b bits long, the above algorithm will try at least $2^{b”}1$ other programs first.

Formal Definitions for P and NP

Conceptually a *decision problem* is a problem that takes as input some string w over an alphabet Σ, and outputs "yes" or "no". If there is an algorithm (say a Turing machine, or a computer program with unbounded memory) that can produce the correct answer for any input string of length n in at most $c \cdot n^k + c$ steps, where k and c are constants independent of the input string, then we say that the problem can be solved in *polynomial time* and we place it in the class P. Formally, P is defined as the set of all languages that can be decided by a deterministic polynomial-time Turing machine. That is,

$$\mathbf{P} = \{L : L = L(M) \text{ for some deterministic polynomial-time Turing machine } M\}$$

where

$$L(M) = \{w \in \Sigma^* : M \text{ accepts } w\}$$

and a deterministic polynomial-time Turing machine is a deterministic Turing machine M that satisfies the following two conditions:

1. M halts on all input w and
2. there exists $k \in N$ such that $T_M(n) \in O(n^k)$ where O refers to the big O notation),

where $T_M(n) = \max\{t_M(w) : w \in \Sigma^*, |w| = n\}$

and $t_M(w)$ = number of steps M takes to halt on input w.

NP can be defined similarly using nondeterministic Turing machines (the traditional way). However, a modern approach to define NP is to use the concept of *certificate* and *verifier*. Formally, NP is defined as the set of languages over a finite alphabet that have a verifier that runs in polynomial time, where the notion of "verifier" is defined as follows.

Let L be a language over a finite alphabet, Σ.

L " NP if, and only if, there exists a binary relation $R \subset \Sigma^* \times \Sigma^*$ and a positive integer k such that the following two conditions are satisfied:

1. For all $x \in \Sigma^*$, $x \in L \Leftrightarrow \exists y \in \Sigma^*$ such that $(x,y) \in R$ and $|y| \in O(|x|^k)$; and
2. the language $L_R = \{x\#y : (x,y) \in R\}$ over $\Sigma \cup \{\#\}$ is decidable by a Turing machine in polynomial time.

A Turing machine that decides L_R is called a *verifier* for L and a y such that is called a*certificate of membership* of x in L. In general, a verifier does not have to be polynomial-time. However, for L to be in NP, there must be a verifier that runs in polynomial time.

Example

Let

$$COMPOSITE = \{x \in \mathbb{N} \mid x = pq \text{ for integers } p,q > 1\}$$

$$R = \left\{(x,y) \in \mathbb{N} \times \mathbb{N} \mid 1 < y \le \sqrt{x} \text{ and } y \text{ divides } x\right\}.$$

Clearly, the question of whether a given x is a composite is equivalent to the question of whether x is a member of COMPOSITE. It can be shown that COMPOSITE " NP by verifying that it satisfies the above definition (if we identify natural numbers with their binary representations).

COMPOSITE also happens to be in P.

Formal Definition for NP-completeness

There are many equivalent ways of describing NP-completeness.

Let L be a language over a finite alphabet Σ.

L is NP-complete if, and only if, the following two conditions are satisfied:

1. $L \in \mathbf{NP}$; and
2. any $L \in \mathbf{NP}$ is polynomial-time-reducible to $\boldsymbol{L}$ (written as $L \leq_p L$), where if, and only if, the following two conditions are satisfied:
1. There exists $f : \Sigma^* \rightarrow \Sigma^*$ such that $\forall w \in \Sigma^* (w \in L \Leftrightarrow f(w) \in L)$; and
2. there exists a polynomial-time Turing machine that halts with on its tape on any input .

The complexity class P is often seen as a mathematical abstraction modeling those computational tasks that admit an efficient algorithm. This hypothesis is called the Cobham–Edmonds thesis.

The complexity class NP, on the other hand, contains many problems that people would like to solve efficiently, but for which no efficient algorithm is known, such as the Boolean satisfiability problem, theHamiltonian path problem and the vertex cover problem. Since deterministic Turing machines are special nondeterministic Turing machines, it is easily observed that each problem in P is also member of the class NP.

The question of whether P equals NP is one of the most important open questions in theoretical computer science because of the wide implications of a solution.

If the answer is yes, many important problems can be shown to have more efficient solutions. These include various types ofinteger programming problems in operations research, many problems in logistics, protein structure prediction in biology, and the ability to find formal proofs of pure mathematics theorems. The P versus NP problem is one of the Millennium Prize Problems proposed by the Clay Mathematics Institute. There is a US$1,000,000 prize for resolving the problem.

Problems in NP not Known to be in P or NP-complete

It was shown by Ladner that if P $\neq$ NP then there exist problems in NP that are neither in P nor NP-complete. Such problems are called NP-intermediate problems. The graph isomorphism problem, the discrete logarithm problem and the integer factorization problem are examples of problems believed to be NP-intermediate. They are some of the very few NP problems not known to be in P or to be NP-complete.

The graph isomorphism problem is the computational problem of determining whether two finite graphsare isomorphic. An important unsolved problem in complexity theory is whether the graph isomorphism problem is in P, NP-complete, or NP-intermediate. The answer is not known, but it is believed that the problem is at least not NP-complete.

If graph isomorphism is NP-complete, the polynomial time hierarchy collapses to its second level. Since it is widely believed that the polynomial hierarchy does not collapse to any finite level, it is believed that graph isomorphism is not NP-complete. The best algorithm for this problem, due to Laszlo Babai and Eugene Luks has run time $2^{O(``(n \log(n)))}$ for graphs with n vertices.

The integer factorization problem is the computational problem of determining the prime factorization of a given integer. Phrased as a decision problem, it is the problem of deciding whether the input has a factor less than k. No efficient integer factorization algorithm is known, and this fact forms the basis of several modern cryptographic systems, such as the RSA algorithm. The integer factorization problem is in NP and in co-NP (and even in UP and co-UP). If the problem is NP-complete, the polynomial time hierarchy will collapse to its first level (i.e., NP will equal co-NP). The best known algorithm for integer factorization is the general number field sieve, which takes time $O(e^{(64/9)1/3(n.\log 2)1/3(\log (n.\log 2))2/3})$ to factor an n-bit integer. However, the best known quantum algorithm for this problem, Shor's algorithm, does run in polynomial time. Unfortunately, this fact doesn't say much about where the problem lies with respect to non-quantum complexity classes.

Separations Between other Complexity Classes

Many known complexity classes are suspected to be unequal, but this has not been proved. For instance $\mathbf{P \subseteq NP \subseteq PP \subseteq PSPACE}$

, but it is possible that P = PSPACE. If P is not equal to NP, thenP is not equal to PSPACE either. Since there are many known complexity classes between P andPSPACE, such as RP, BPP, PP, BQP, MA, PH, etc., it is possible that all these complexity classes collapse to one class. Proving that any of these classes are unequal would be a major breakthrough in complexity theory.

Along the same lines, co-NP is the class containing the complement problems (i.e. problems with the*yes*/*no* answers reversed) of NP problems. It is believed that NP is not equal to co-NP; however, it has not yet been proven. It has been shown that if these two complexity classes are not equal then Pis not equal to NP. Similarly, it is not known if L (the set of all problems that can be solved in logarithmic space) is strictly contained in P or equal to P. Again, there are many complexity classes between the two, such as NLand NC, and it is not known if they are distinct or equal classes.

It is suspected that P and BPP are equal. However, it is currently open if BPP = NEXP.

Intractability

Problems that can be solved in theory (e.g., given infinite time), but which in practice take too long for their solutions to be useful, are known as *intractable* problems. In complexity theory, problems that lack polynomial-time solutions are considered to be intractable for more than the smallest inputs.

In fact, the Cobham–Edmonds thesis states that only those problems that can be solved in polynomial time can be feasibly computed on some computational device. Problems that are known to be intractable in this sense include those that are EXPTIME-hard. If NP is not the same as P, then the NP-complete problems are also intractable in this sense.

To see why exponential-time algorithms might be unusable in practice, consider a program that makes 2^n operations before halting. For small$n$, say 100, and assuming for the sake of example that the computer does 10^{12} operations each second, the program would run for about 4×10^{10} years, which is roughly the age of the universe. Even with a much faster computer, the program would only be useful for very small instances and in that sense the intractability of a problem is somewhat independent of technological progress. Nevertheless a polynomial time algorithm is not always practical. If its running time is, say, n^{15}, it is unreasonable to consider it efficient and it is still useless except on small instances.

What intractability means in practice is open to debate. Saying that a problem is not in P does not imply that all large cases of the problem are hard or even that most of them are. For example the decision problem in Presburger arithmetic has been shown not to be in P, yet algorithms have been written that solve the problem in reasonable times in most cases. Similarly, algorithms can solve the

NP-complete knapsack problem over a wide range of sizes in less than quadratic time and SAT solvers routinely handle large instances of the NP-complete Boolean satisfiability problem.

Continuous Complexity Theory

Continuous complexity theory can refer to complexity theory of problems that involve continuous functions that are approximated by discretizations, as studied in numerical analysis. One approach to complexity theory of numerical analysis is information based complexity.

Continuous complexity theory can also refer to complexity theory of the use of analog computation, which uses continuous dynamical systems and differential equations. Control theory can be considered a form of computation and differential equations are used in the modelling of continuous-time and hybrid discrete-continuous-time systems.

History

Before the actual research explicitly devoted to the complexity of algorithmic problems started off, numerous foundations were laid out by various researchers. Most influential among these was the definition of Turing machines by Alan Turing in 1936, which turned out to be a very robust and flexible notion of computer.

Fortnow & Homer (2003) date the beginning of systematic studies in computational complexity to the seminal paper "On the Computational Complexity of Algorithms" by Juris Hartmanis and Richard Stearns (1965), which laid out the definitions of time and space complexity and proved the hierarchy theorems.

According to Fortnow & Homer (2003), earlier papers studying problems solvable by Turing machines with specific bounded resources include John Myhill's definition of linear bounded automata (Myhill 1960), Raymond Smullyan's study of rudimentary sets (1961), as well as Hisao Yamada's paper on real-time computations (1962). Somewhat earlier, Boris Trakhtenbrot (1956), a pioneer in the field from the USSR, studied another specific complexity measure. As he remembers:

However, [my] initial interest [in automata theory] was increasingly set aside in favor of computational complexity, an exciting fusion of combinatorial methods, inherited from switching theory, with the conceptual arsenal of the theory of algorithms. These ideas had occurred to me earlier in 1955 when I coined the term "signalizing function", which is nowadays commonly known as "complexity measure". —Boris Trakhtenbrot, *From Logic to Theoretical Computer Science – An Update. In: Pillars of Computer Science, LNCS 4800, Springer 2008.*

In 1967, Manuel Blum developed an axiomatic complexity theory based on his axioms and proved an important result, the so called, speed-up theorem. The field really began to flourish when the US researcher Stephen Cook and, working independently, Leonid Levin in the USSR, proved that there exist practically relevant problems that are NP-complete. In 1972, Richard Karp took this idea a leap forward with his landmark paper, "Reducibility Among Combinatorial Problems", in which he showed that 21 diverse combinatorial and graph theoretical problems, each infamous for its computational intractability, are NP-complete.

Chapter 8

Complexity Classes

This is a list of complexity classes in computational complexity theory. For other computational and complexity subjects, see list of computability and complexity topics.

Many of these classes have a 'Co' partner which consists of the complements of all languages in the original class. For example if a language L is in NP then the complement of L is in Co-NP. (This doesn't mean that the complement of NP is Co-NP - there are languages which are known to be in both, and other languages which are known to be in neither.)

"The hardest problems" of a class refer to problems, which belong to the class and every other problem of that class can be reduced to it. Furthermore, the reduction is also a problem of the given class, or its subset.

If you don't see a class listed (such as Co-UP) you should look under its partner (such as UP).

Sharp-P

The correct title of this article is #P. The substitution or omission of the # sign is because oftechnical restrictions.

In computational complexity theory, the complexity class #P (pronounced "number P" or, sometimes "sharp P" or "hash P") is the set of the counting problems associated with the decision problems in the set NP.

More formally, #P is the class of function problems of the form "compute $f(x)$," where f is the number of accepting paths of a nondeterministic Turing machine running in polynomial time. Unlike most well-known complexity classes, it is not a class of decision problems but a class of function problems.

An NP problem is often of the form, "Are there any solutions that satisfy certain constraints?" For example:

- Are there any subsets of a list of integers that add up to zero? (subset sum problem)
- Are there any Hamiltonian cycles in a given graph with cost less than 100? (traveling salesman problem)
- Are there any variable assignments that satisfy a given CNF formula? (Boolean satisfiability problem)

The corresponding #P problems ask "how many" rather than "are there any". For example:

- How many subsets of a list of integers add up to zero?
- How many Hamiltonian cycles in a given graph have cost less than 100?
- How many variable assignments satisfy a given CNF formula?

Clearly, a #P problem must be at least as hard as the corresponding NP problem. If it's easy to count answers, then it must be easy to tell whether there are any answers. Just count them, and see if the count is greater than zero.

One consequence of Toda's theorem is that a polynomial-time machine with a #P oracle ($P^{\#P}$) can solve all problems in PH, the entire polynomial hierarchy. In fact, the polynomial-time machine only needs to make one #P query to solve any problem in PH. This is an indication of the extreme difficulty of solving #P-complete problems exactly.

Surprisingly, some #P problems that are believed to be difficult correspond to easy P problems. For more information on this, see #P-complete.

The closest decision problem class to #P is PP, which asks if a majority (more than half) of the computation paths accept. This finds the most significant bit in the #P problem answer. The decision problem class •"P instead asks for the least significant bit of the #P answer.

The complexity class #P was first defined by Leslie Valiant in a 1979 paper on the computation of thepermanent, in which he proved that permanent is #P-complete.

Larry Stockmeyer has proved that for every #P problem *P* there exists a randomized algorithm using oracle for SAT, which given an instance *a* of *P* and $\varepsilon>0$ returns with high probability a number *x* such that $(1-\epsilon)P(a) \leq x \leq (1+\epsilon)P(a)$. The runtime of the algorithm is polynomial in a and $1/\varepsilon$. The algorithm is based on leftover hash lemma.

Sharp-P-complete

The correct title of this article is #P-complete. The substitution or omission of the # sign is because of technical restrictions.

#P-complete, pronounced "sharp P complete" or "number P complete" is a complexity class incomputational complexity theory. A problem is #P-complete if and only if it is in #P, and every problem in #P can be reduced to it by a polynomial-time counting reduction, i.e. a polynomial-timeTuring reduction relating the cardinalities of solution sets.▯ Equivalently, a problem is #P-complete if and only if it is in #P, and for any non-deterministic Turing machine ("NP machine"), the problem of computing its number of accepting paths can be reduced to this problem.▯

Very often the reductions are "parsimonious," i.e., they preserve the number of solutions.▯

Examples of #P-complete problems include:

- How many different variable assignments will satisfy a given general boolean formula? (#SAT)
- How many different variable assignments will satisfy a given DNF formula?
- How many different variable assignments will satisfy a given 2SAT formula?
- How many perfect matchings are there for a given bipartite graph?
- What is the value of the permanent of a given matrix whose entries are 0 or 1? (See Permanent is sharp-P-complete.)
- How many graph colorings using k colors are there for a particular graph G?
- How many different linear extensions are there for a given partial order, or, equivalently, how many different topological orderings are there for a given directed acyclic graph?

A polynomial-time algorithm for solving a #P-complete problem, if it existed, would imply P = NP, and thus P = PH. No such algorithm is currently known.

Easy Problems with Hard Counting Versions

It is surprising that some #P-complete problems correspond to easy P problems. Its very easy to determine the satisfiability of a boolean formula in DNF: such a formula is satisfiable if and only if it contains a satisfiable conjunction (one that does not contain a variable and its negation), whereas counting the number of satisfying assignments is #P-complete. Also deciding 2-SAT is easy in contrast to counting the number of satisfying assignments. Topologically sorting is easy in contrast to counting the number of topological sortings.

The same observation can be made for the perfect matching problem. It was known before that the decision problem “Is there a perfect matching for a given bipartite graph?” can be solved in polynomial time, and in fact, for a graph with V vertices and Eedges, it can be solved in $O(VE)$ time.

The corresponding question “How many perfect matchings does the given bipartite graph have?” is already #P-complete. The problem of counting the number of perfect matchings (or in directed graphs: the number of vertex cycle covers) is known to be equivalent to the problem of the computation of the permanent of a matrix. The perfect matching counting problem was the first counting problem corresponding to an easy P problem shown to be #P-complete, in a 1979 paper by Leslie Valiant which also defined the classes #P and #P-complete for the first time.

Approximation

There are probabilistic algorithms that return good approximations to some #P-complete problems with high probability. This is one of the demonstrations of the power of probabilistic algorithms.

Many #P-complete problems have a fully polynomial-time randomized approximation scheme, or “FPRAS,” which, informally, will produce with high probability an approximation to an arbitrary degree of accuracy, in time that is polynomial with respect to both the size of the problem and the degree of accuracy required. Jerrum, Valiant, and Vazirani showed that every #P-complete problem either has an FPRAS, or is essentially impossible to approximate; if there is any polynomial-time algorithm which consistently produces an approximation of a #P-complete problem which is within a polynomial ratio in the size of the input of the exact answer, then that algorithm can be used to construct an FPRAS.

2-EXPTIME

In computational complexity theory, the complexity class 2-EXPTIME (sometimes called 2-EXP) is the set of all decision problems solvable by a deterministic Turing machine in $O(2^{2^{p(n)}})$ time, where $p(n)$ is a polynomial function of n.

In terms of DTIME,

$$\text{2-EXPTIME} = \bigcup_{k \in \mathbb{N}} \text{DTIME}\left(2^{2^{n^k}}\right).$$

We know

$\text{P} \subseteq \text{NP} \subseteq \text{PSPACE} \subseteq \text{EXPTIME} \subseteq \text{NEXPTIME} \subseteq \text{EXPSPACE} \subseteq \text{2-EXPTIME} \subseteq \text{ELEMENTARY}$.

2-EXPTIME can also be reformulated as the space class AEXPSPACE, the problems that can be solved by an alternating Turing machine in exponential space. This is one way to see that EXPSPACE 2-$\subseteq$ EXPTIME, since an alternating Turing machine is at least as powerful as a deterministic Turing machine.

2-EXPTIME is one class in a hierarchy of complexity classes with increasingly higher time bounds. The class 3-EXPTIME is defined similarly to 2-EXPTIME but with a triply exponential time bound $2^{2^{2^{n^k}}}$. This can be generalized to higher and higher time bounds.

2-EXPTIME-complete Problems

Generalizations of many fully observable games are EXPTIME-complete. These games can be viewed as particular instance of a class of transition systems defined in terms of a set of state variables and actions/events that change the values of the state variables, together with the question of whether a winning strategy exists.

A generalization of this class of fully observable problems to partially observable problems lifts the complexity from EXPTIME-complete to 2-EXPTIME-complete.

AC^0

AC^0 is a complexity class used in circuit complexity. It is the smallest class in the AC hierarchy, and consists of all families of circuits of depth O(1) and polynomial size, with unlimited-fanin AND gatesand OR gates. (We allow NOT gates only at the inputs). It thus contains NC^0, which has only bounded-fanin AND and OR gates.

From a descriptive complexity viewpoint, DLOGTIME-uniform AC^0 is equal to the descriptive classFO+BIT of all languages describable in first-order logic with the addition of the BIT operator, or alternatively by FO(+, ×), or by Turing machine in the logarithmic hierarchy.

In 1984 Furst, Saxe, and Sipser showed that calculating the parity of an input cannot be decided by any AC^0 circuits, even with non-uniformity. It follows that AC^0 is not equal to NC^1, because a family of circuits in the latter class can compute parity.

ACC^0

ACC^0, sometimes called ACC, is a class of computational models and problems defined in circuit complexity, a field of theoretical computer science. The class is defined as by augmenting the classAC^0 of "alternating circuits" with the ability to count; the acronym ACC stands for "AC with counters".

Specifically, a problem belongs ACC^0 if it can be solved by polynomial-size, constant-depth circuits of unbounded fan-in gates, including gates that count modulo a fixed integer. ACC^0corresponds to computation in any solvable monoid. The class is very well studied in theoretical computer science because of the algebraic connections and because it is one of the largest concrete computational models for which computational impossibility results, so-called circuit lower bounds, can be proved.

Definitions

Informally, ACC^0 models the class of computations realised by Boolean circuits of constant depth and polynomial size, where the circuit gates includes "modular counting gates" that compute the number of true inputs modulo some fixed constant. More formally, a language belongs to $AC^0[m]$ if it can be computed by a family of circuits C_1, C_2, ..., where C_n takes n inputs, the depth of every circuit is constant, the size of C_n is a polynomial function of n, and the circuit uses the following gates: AND gates and OR gates of unbounded fan-in, computing the conjunction and disjunction of their inputs; NOT gates computing the negation of their single input; and unbounded fan-in MOD-m gates, which compute 1 if the number of input 1s is a multiple of m. A language belongs to ACC^0 if it belongs to $AC^0[m]$ for some m. In some texts, ACC^i refers to a hierarchy of circuit classes with ACC^0 at its lowest level, where the circuits in ACC^i have depth $O(\log^i n)$ and polynomial size.

The class ACC^0 can also be defined in terms of computations of nonuniform deterministic finite automata (NUDFA) over monoids. In this framework, the input is interpreted as elements from a fixed monoid, and the input is accepted if the product of the input elements belongs to a given list of monoid elements.

The class ACC^0 is the family of languages accepted by a NUDFA over some monoid that does not contain an unsolvable group as a subsemigroup.

Computational Power

The class ACC^0 includes AC^0. This inclusion is strict, because a single MOD-2 gate computes the parity function, which is known to be impossible in AC^0. More generally, the function MOD_m can not be computed in $AC^0[p]$ for prime p unless m is a power of p.

Every problem in ACC^0 can be solved by circuits of depth 2, with AND gates of polylogarithmic fan-in at the inputs, connected to a single gate computing a symmetric function. These circuits are called SYM^+-circuits.

The class ACC^0 is included in TC^0.

A 2010 manuscript of Ryan Williams shows that ACC^0 does not contain NEXPTIME. The proof bases on many results in complexity theory, including time hierarchy theorem, IP = PSPACE,derandomization, ideas in proof of Toda's theorem.

It is known that computing the permanent is impossible for logspace-uniform ACC^0 circuits, which implies that the complexity class PP is not contained in logspace-uniform ACC^0.

It is conjectured that ACC^0 is unable to compute the majority function of its inputs, but this remains unresolved as of 2010.

AC (Complexity)

In circuit complexity, AC is a complexity class hierarchy. Each class, AC^i, consists of the languagesrecognized by Boolean circuits with depth $O(\log^i n)$ and a polynomial number of unlimited-fanin ANDand OR gates.

The name "AC" was chosen by analogy to NC, with the "A" in the name standing for "alternating" and referring both to the alternation between the AND and OR gates in the circuits and to alternating Turing machines.

The smallest AC class is AC^0, consisting of constant-depth unlimited-fanin circuits.

The total hierarchy of AC classes is defined as

$$AC = \bigcup_{i \geq 0} AC^i$$

Relation to NC

The AC classes are related to the NC classes, which are defined similarly, but with gates having only constant fanin. For each i, we have

$$NC^i \subseteq AC^i \subseteq NC^{i+1}.$$

As an immediate consequence of this, we have that NC = AC.

Variations

The power of the AC classes can be affected by adding additional gates. If we add gates which calculate the modulo operation for some modulus m, we have the classes $ACC^i[m]$.

Arithmetical Hierarchy

In mathematical logic, the arithmetical hierarchy, arithmetic hierarchy or Kleene-Mostowski hierarchy classifies certain sets based on the complexity of formulas that define them. Any set that receives a classification is called arithmetical.

The arithmetical hierarchy is important in recursion theory, effective descriptive set theory, and the study of formal theories such as Peano arithmetic.

The Tarski-Kuratowski algorithm provides an easy way to get an upper bound on the classifications assigned to a formula and the set it defines.

The hyperarithmetical hierarchy and the analytical hierarchy extend the arithmetical hierarchy to classify additional formulas and sets.

The Arithmetical Hierarchy of Formulas

The arithmetical hierarchy assigns classifications to the formulas in the language of first-order arithmetic. The classifications are denoted Σ_n^0 and Π_n^0 for natural numbers n (including 0). The Greek letters here are lightface symbols, which indicates that the formulas do not contain set parameters.

If a formula ϕ is logically equivalent to a formula with only bounded quantifiers then ϕ is assigned the classifications Σ_0^0 and Π_0^0.

The classifications and are defined inductively for every natural number n using the following rules:

- If ϕ is logically equivalent to a formula of the form $\exists n_1 \exists n_2 \cdots \exists n_k \psi$, where ψ is Π_n^0, then $\phi$is assigned the classification Σ_{n+1}^0.
- If ϕ is logically equivalent to a formula of the form $\forall n_1 \forall n_2 \cdots \forall n_k \psi$ where ψ is Σ_n^0, then $\phi$is assigned the classification Π_{n+1}^0.

As well, a Σ_n^0 formula is equivalent to a formula that begins with some existential quantifiers and alternates n " 1 times between series of existential and universal quantifiers; while a Π_n^0 formula is equivalent to a formula that begins with some universal quantifiers and alternates similarly.

Because every formula is equivalent to a formula in prenex normal form, every formula with no set quantifiers is assigned at least one classification. Because redundant quantifiers can be added to any formula, once a formula is assigned the classification Σ_n^0 or Π_n^0 it will be assigned the classifications Σ_m^0 and Π_m^0 for every m greater than n. The most important classification assigned to a formula is thus the one with the least n, because this is enough to determine all the other classifications.

The Arithmetical Hierarchy of Sets of Natural Numbers

A set X of natural numbers is defined by formula φ in the language of Peano arithmetic if the elements of X are exactly the numbers that satisfy φ. That is, for all natural numbers n,

$$n \in X \Leftrightarrow \mathbb{N} \vDash \phi(\underline{n}),$$

where $\underline{n}$ is the numeral in the language of arithmetic corresponding to n. A set is definable in first order arithmetic if it is defined by some formula in the language of Peano arithmetic.

Each set X of natural numbers that is definable in first order arithmetic is assigned classifications of the form Σ_n^0, Π_n^0, and Δ_n^0, where n is a natural number, as follows. If X is definable by a Σ_n^0 formula then X is assigned the classification Σ_n^0 . If X is definable by a Π_n^0 formula then X is assigned the classification Π_n^0 . If X is both Σ_n^0 and Σ_n^0 then X is assigned the additional classification Δ_n^0 .

Note that it rarely makes sense to speak of Δ_n^0 formulas; the first quantifier of a formula is either existential or universal. So Δ_n^0 a set is not defined by Δ_n^0 a formula; rather, there are both Σ_n^0 and Π_n^0 formulas that define the set.

A parallel definition is used to define the arithmetical hierarchy on finite Cartesian powers of the natural numbers. Instead of formulas with one free variable, formulas with k free number variables are used to define the arithmetical hierarchy on sets of k-tuples of natural numbers.

Relativized Arithmetical Hierarchies

Just as we can define what it means for a set X to be recursive relative to another set Y by allowing the computation defining X to consult Y as an oracle we can extend this notion to the whole arithmetic hierarchy and define what it means for X to be Σ_n^0, Δ_n^0 or Π_n^0 in Y, denoted respectively $\Sigma_n^{0,Y}$ $\Delta_n^{0,Y}$ and $\Pi_n^{0,Y}$. To do so, fix a set of integers Y and add a predicate for membership in Y to the language of Peano arithmetic. We then say that X is in if it is defined by a Σ_n^0 formula in this expanded language. In other words X $\Sigma_n^{0,Y}$ is if it is defined by a Σ_n^0 formula allowed to ask questions about membership in Y. Alternatively one can view the $\Sigma_n^{0,Y}$ sets as those sets that can be built starting with sets recursive in Y and alternatively projecting and taking the complements of these sets up to n times.

For example let Y be a set of integers. Let X be the set of numbers divisible by an element of Y. Then$X$ is defined by the formula $\phi(n) = \exists m \exists t(Y(m) \wedge m \times t = n)$ so X is in $\Sigma_1^{0,Y}$ (actually it is in $\Delta_0^{0,Y}$ as well since we could bound both quantifiers by n).

Arithmetic Reducibility and Degrees

Arithmetical reducibility is an intermediate notion between Turing reducibility and hyperarithmetic reducibility.

A set is arithmetical (also arithmetic and arithmetically definable) if it is defined by some formula in the language of Peano arithmetic. Equivalently X is arithmetical if X is Σ_n^0 or Π_n^0 for some integer$n$. A set X is arithmetical in a set Y, denoted $X \leq_A Y$, if X is definable a some formula in the language of Peano arithmetic extended by a predicate for membership in Y. Equivalently, X is arithmetical in Y if X is in $\Sigma_n^{0,Y}$ or $\Pi_n^{0,Y}$ for some integer n. A synonym for is: X isarithmetically reducible to Y.

The relation $X \leq_A Y$ is reflexive and transitive, and thus the relation $\equiv_A$ defined by the rule

$$X \equiv_A Y \Leftrightarrow X \leq_A Y \wedge Y \leq_A X$$

is an equivalence relation. The equivalence classes of this relation are called the arithmetic degrees; they are partially ordered under $\leq_A$.

The Arithmetical Hierarchy of Subsets of Cantor and Baire Space

The Cantor space, denoted 2^ω, is the set of all infinite sequences of 0s and 1s; the Baire space, denoted ω^ω or $\mathcal{N}$, is the set of all infinite sequences of natural numbers. Note that elements of the Cantor space can be identified with sets of integers and elements of the Baire space with functions from integers to integers.

The ordinary axiomatization of second-order arithmetic uses a set-based language in which the set quantifiers can naturally be viewed as quantifying over Cantor space. A subset of Cantor space is assigned the classification Σ_n^0 if it is definable by a formula. The set is assigned the classification if it is definable by a formula.

If the set is both and then it is given the additional classification . For example let be the set of all infinite binary strings which aren't all 0 (or equivalently the set of all non-empty sets of integers). As we see that O is defined by a formula and hence is a set.

Note that while both the elements of the Cantor space (regarded as sets of integers) and subsets of the Cantor space are classified in arithmetic hierarchies but these are not the same hierarchy. In fact the relationship between the two hierarchies is interesting and non-trivial.

For instance the Π_n^0 elements of the Cantor space are not (in general) the same as the elements X of the Cantor space so that $\{X\}$ is a subset of the Cantor space. However, many interesting results relate the two hierarchies.

There are two ways that a subset of Baire space can be classified in the arithmetical hierarchy.

- A subset of Baire space has a corresponding subset of Cantor space under the map that takes each function from ω to ω to the characteristic function of its graph. A subset of Baire space is given the classification Σ^1_n, Π^1_n, Δ^1_n or if and only if the corresponding subset of Cantor space has the same classification.
- An equivalent definition of the analytical hierarchy on Baire space is given by defining the analytical hierarchy of formulas using a functional version of second-order arithmetic; then the analytical hierarchy on subsets of Cantor space can be defined from the hierarchy on Baire space. This alternate definition gives exactly the same classifications as the first definition.

A parallel definition is used to define the arithmetical hierarchy on finite Cartesian powers of Baire space or Cantor space, using formulas with several free variables. The arithmetical hierarchy can be defined on any effective Polish space; the definition is particularly simple for Cantor space and Baire space because they fit with the language of ordinary second-order arithmetic.

Note that we can also define the arithmetic hierarchy of subsets of the Cantor and Baire spaces relative to some set of integers. In fact boldface Σ^0_n is just the union of $\Sigma^{0,Y}_n$ for all sets of integers Y. Note that the boldface hierarchy is just the standard hierarchy of Borel sets.

Extensions and Variations

It is possible to define the arithmetical hierarchy of formulas using a language extended with a function symbol for each primitive recursive function. This variation slightly changes the classification of some sets. A more semantic variation of the hierarchy can be defined on all finitary relations on the natural numbers; the following definition is used. Every computable relation is defined to be Σ^0_0 and Π^0_0. The classifications Σ^0_n and Π^0_n are defined inductively with the following rules.

- If the relation $R(n_1,\ldots,n_l,m_1,\ldots,m_k)$ is Σ^0_n then the $S(n_1,\ldots,n_l) = \forall m_1 \cdots \forall m_k R(n_1,\ldots,n_l,m_1,\ldots,m_k)$ is defined to be Π^0_{n+1}
- If the relation $R(n_1,\ldots,n_l,m_1,\ldots,m_k)$ is Π^0_n then the relation $S(n_1,\ldots,n_l) = \exists m_1 \cdots \exists m_k R(n_1,\ldots,n_l,m_1,\ldots,m_k)$ is defined to be Σ^0_{n+1}

This variation slightly changes the classification of some sets. It can be extended to cover finitary relations on the natural numbers, Baire space, and Cantor space.

Meaning of the Notation

The following meanings can be attached to the notation for the arithmetical hierarchy on formulas.

The subscript n in the symbols Σ_n^0 and Π_n^0 indicates the number of alternations of blocks of universal and existential number quantifiers that are used in a formula. Moreover, the outermost block is existential in Σ_n^0 formulas and universal in Π_n^0 formulas.

The superscript 0 in the symbols Σ_n^0, Π_n^0, Δ_n^0 and indicates the type of the objects being quantified over. Type 0 objects are natural numbers, and objects of type $i + 1$ are functions that map the set of objects of type i to the natural numbers. Quantification over higher type objects, such as functions from natural numbers to natural numbers, is described by a superscript greater than 0, as in theanalytical hierarchy.

The superscript 0 indicates quantifiers over numbers, the superscript 1 would indicate quantification over functions from numbers to numbers (type 1 objects), the superscript 2 would correspond to quantification over functions that take a type 1 object and return a number, and so on.

Examples

- The Σ_1^0 sets of numbers are those definable by a formula of the form $\exists n_1 \cdots \exists n_k \psi(n_1,\ldots,n_k,m)$ where ψ has only bounded quantifiers. These are exactly the recursively enumerable sets.
- The set of natural numbers that are indices for Turing machines that compute total functions is Π_2^0. Intuitively, an index e falls into this set if and only if for every m "there is an s such that the Turing machine with index e halts on input m after s steps". A complete proof would show that the property displayed in quotes in the previous sentence is definable in the language of Peano arithmetic by a formula.
- Every Σ_1^0 subset of Baire space or Cantor space is an open set in the usual topology on the space. Moreover, for any such set there is a computable enumeration of Gödel numbers of basic

open sets whose union is the original set. For this reason, Σ_1^0 sets are sometimes called*effectively open*. Similarly, every Π_1^0 set is closed and the Π_1^0 sets are sometimes called*effectively closed*.

- Every arithmetical subset of Cantor space of(or?) Baire space is a Borel set. The lightface Borel hierarchy extends the arithmetical hierarchy to include additional Borel sets. For example, every Π_2^0 subset of Cantor or Baire space is a G_δ set (that is, a set which equals the intersection of countably many open sets). Moreover, each of these open sets is Σ_1^0 and the list of Gödel numbers of these open sets has a computable enumeration. If $\phi(X,n,m)$ is a Σ_0^0 formula with a free set variable X and free number variables n,m then the Π_2^0 set $\{X \mid \forall n \exists m \phi(X,n,m)\}$ is the intersection of the Σ_1^0 sets of the form $\{X \mid \exists m \phi(X,n,m)\}$ as n ranges over the set of natural numbers.

Properties

The following properties hold for the arithmetical hierarchy of sets of natural numbers and the arithmetical hierarchy of subsets of Cantor or Baire space.

- The collections Π_n^0 and Σ_n^0 are closed under finite unions and finite intersections of their respective elements.
- A set is Σ_n^0 if and only if its complement is Π_n^0. A set is Δ_n^0 if and only if the set is both Σ_n^0 and, Π_n^0 in which case its complement will also be .
- The inclusions $\Delta_n^0 \subsetneq \Pi_n^0$ and $\Delta_n^0 \subsetneq \Sigma_n^0$ hold for $n \geq 1$.
- The inclusions $\Pi_n^0 \subsetneq \Pi_{n+1}^0$ and $\Sigma_n^0 \subsetneq \Sigma_{n+1}^0$ hold for all n and the inclusion $\Sigma_n^0 \cup \Pi_n^0 \subsetneq \Delta_{n+1}^0$ holds $n \geq 1$ for . Thus the hierarchy does not collapse.

Bounded-error Probabilistic Polynomial

In computational complexity theory, bounded-error probabilistic polynomial time (BPP) is the class of decision problems solvable by a probabilistic Turing machine in polynomial time, with an errorprobability of at most 1/3 for all instances

BPP algorithm (1 run)		
	Answer produced	
Correct answer	YES	NO
YES	? 2/3	? 1/3
NO	? 1/3	? 2/3
BPP algorithm (*k* runs)		
	Answer produced	
Correct answer	YES	NO
YES	> 1 ? $2^{?ck}$	$< 2^{?ck}$
NO	$< 2^{?ck}$	> 1 ? $2^{?ck}$
for some constant $c > 0$		

Informally, a problem is in BPP if there is an algorithm for it that has the following properties:

- It is allowed to flip coins and make random decisions
- It is guaranteed to run in polynomial time
- On any given run of the algorithm, it has a probability of at most 1/3 of giving the wrong answer, whether the answer is YES or NO.

BPP is one of the largest *practical* classes of problems, meaning most problems of interest in BPP have efficient probabilistic algorithms that can be run quickly on real modern machines. For this reason it is of great practical interest which problems and classes of problems are inside BPP. BPP also contains P, the class of problems solvable in polynomial time with a deterministic machine, since a deterministic machine is a special case of a probabilistic machine.

In practice, an error probability of 1/3 might not be acceptable, however, the choice of 1/3 in the definition is arbitrary. It can be any constant between 0 and 1/2 (exclusive) and the set BPP will be unchanged. It does not even have to be constant: the same class of problems is defined by allowing error as high as 1/2 “ $n^{”c}$ on the one hand, or requiring error as small as 2^{-nc} on the other hand, where c is any positive constant, and n is the length of input. The idea is that there is a probability of error, but if the algorithm is run many times, the chance that the majority of the runs are wrong drops off exponentially as a consequence of the Chernoff bound. This makes it possible to create a highly accurate algorithm by merely running the algorithm several times and taking a “majority vote” of the answers. For example, if one defined the class with the restriction that the algorithm can be wrong with probability at most $1/2^{100}$, this would result in the same class of problems.

Definition

A language L is in BPP if and only if there exists a probabilistic Turing machine M, such that

- M runs for polynomial time on all inputs
- For all x in L, M outputs 1 with probability greater than or equal to 2/3
- For all x not in L, M outputs 1 with probability less than or equal to 1/3

Unlike the complexity class ZPP, the machine M is required to run for polynomial time on all inputs, regardless of the outcome of the random coin flips.

Alternatively, BPP can be defined using only deterministic Turing machines. A language L is in BPP if and only if there exists a polynomial p and deterministic Turing machine M, such that

- M runs for polynomial time on all inputs
- For all x in L, the fraction of strings y of length $p(|x|)$ which satisfy $M(x,y) = 1$ is greater than or equal to 2/3
- For all x in not in L, the fraction of strings y of length $p(|x|)$ which satisfy $M(x,y) = 1$ is less than or equal to 1/3

In this definition, the string y corresponds to the output of the random coin flips that the probabilistic Turing machine would have made. For some applications this definition is preferable since it does not mention probabilistic Turing machines.

Problems

Besides the problems in P, which are obviously in BPP, many problems were known to be in BPP but not known to be in P. The number of such problems is decreasing, and it is conjectured thatP = BPP.

For a long time, one of the most famous problems that was known to be in BPP but not known to be in P was the problem ofdetermining whether a given number is prime. However, in the 2002 paper *PRIMES is in P*, Manindra Agrawal and his students Neeraj Kayal and Nitin Saxena found a deterministic polynomial-time algorithm for this problem, thus showing that it is in P.

An important example of a problem in BPP (in fact in co-RP) still not known to be in P is polynomial identity testing, the problem of determining whether a polynomial is identically equal to the zero polynomial. In other words, is there an assignment of variables such that when the polynomial is evaluated the result is nonzero? It suffices to choose each variable's value uniformly at random from a finite subset of at least d values to achieve bounded error probability, where d is the total degree of the polynomial.

Related Classes

If the access to randomness is removed from the definition of BPP, we get the complexity class P. In the definition of the class, if we replace the ordinary Turing machine with a quantum computer, we get the class BQP.

A Monte Carlo algorithm is a randomized algorithm which is likely to be correct. Problems in the classBPP have Monte Carlo algorithms with polynomial bounded running time. This is compared to a Las Vegas algorithm which is a randomized algorithm which either outputs the correct answer, or outputs "fail" with low probability. Las Vegas algorithms with polynomial bound running times are used to define the class ZPP. Alternatively, ZPP contains probabilistic algorithms that are always correct and have expected polynomial running time. This is weaker than saying it is a polynomial time algorithm, since it may run for super-polynomial time, but with very low probability.

Complexity-theoretic Properties

It is known that BPP is closed under complement; that is, BPP = co-BPP. BPP is low for itself, meaning that a BPP machine with the power to solve BPP problems instantly (a BPP oracle machine) is not any more powerful than the machine without this extra power. In symbols, BPP^{BPP} = BPP.

The relationship between BPP and NP is unknown: it is not known if BPP is a subset of NP, NP is a subset of BPP or if they are incomparable. If NP is contained in BPP, which is considered unlikely since it would imply practical solutions for NP-complete problems, then NP = RP and PH †" BPP.

It is known that RP is a subset of BPP, and BPP is a subset of PP. It is not known whether those two are strict subsets, since we don't even know if P is a strict subset of PSPACE. BPP is contained in the second level of the polynomial hierarchy and therefore it is contained in PH. More precisely, theSipser–Lautemann theorem states that BPP †" Σ_2)" Π_2. As a result, P = NP leads to P = BPP sincePH collapses to P in this case. Thus either P = BPP or P ≠ NP or both.

Adleman's theorem states that membership in any language in BPP can be determined by a family of polynomial-size Boolean circuits, which means BPP is contained in P/poly. Indeed, as a consequence of the proof of this fact, every BPP algorithm operating on inputs of bounded length can be derandomized into a deterministic algorithm using a fixed string of random bits. Finding this string may be expensive, however.

Relative to oracles, we know that there exist oracles A and B, such that $P^A = BPP^A$ and $P^B \neq BPP^B$. Moreover, relative to a random oracle with probability 1, P = BPP and BPP is strictly contained in NPand co-NP.

Derandomization

The existence of certain strong pseudorandom number generators is conjectured by most experts of the field. This conjecture implies that randomness does not give additional computational power to polynomial time computation, that is, P = RP = BPP. Note that ordinary generators are not sufficient to show this result; any probabilistic algorithm implemented using a typical random number generator will always produce incorrect results on certain inputs irrespective of the seed (though these inputs might be rare). László Babai, Lance Fortnow, Noam Nisan, and Avi Wigderson showed that unless EXPTIMEcollapses to MA, BPP is contained in i.o.-

SUBEXP = $\bigcap_{\varepsilon>0}$i.o.-DTIME($2^{n^{\varepsilon}}$) The class i.o.-SUBEXP, which stands for infinitely often SUBEXP, contains problems which have sub-exponential time algorithms for infinitely many input sizes. They also showed that P = BPP if the exponential-time hierarchy, which is defined in terms of the polynomial hierarchy and E as E^{PH}, collapses to E; however, note that the exponential-time hierarchy is usually conjectured *not* to collapse.

Russell Impagliazzo and Avi Wigderson showed that if any problem in E, where $E = \mathrm{DTIME}\left(2^{O(n)}\right)$, has circuit complexity $2^{\Omega(n)}$ then.

Parameterized Complexity

In computer science, parameterized complexity is a branch of computational complexity theorythat focuses on classifying computational problems according to their inherent difficulty with respect to*multiple* parameters of the input. The complexity of a problem is then measured as a function in those parameters. This allows to classify NP-hard problems on a finer scale than in the classical setting, where the complexity of a problem is only measured by the number of bits in the input. The first systematic work on parameterized complexity was done by Downey & Fellows (1999).

Under the assumption that P ≠ NP, there exist many natural problems that require superpolynomialrunning time when complexity is measured in terms of the input size only, but that are computable in a time that is polynomial in the input size and exponential or worse in a parameter k. Hence, if k is fixed at a small value and the growth of the function over k is relatively small then such problems can still be considered "tractable" despite their traditional classification as "intractable".

The existence of efficient, exact, and deterministic solving algorithms for NP-complete, or otherwiseNP-hard, problems is considered unlikely, if input parameters are not fixed; all known solving algorithms for these problems require time that is exponential in the total size of the input. However, some problems can be solved by algorithms that are exponential only in the size of a fixed parameter while polynomial in the size of the input. Such an algorithm is called a fixed-parameter tractable (fpt-)algorithm, because the problem can be solved efficiently for small values of the fixed parameter.

Problems in which some parameter k is fixed are called parameterized problems. A parameterized problem that allows for such an fpt-algorithm is said to be a fixed-parameter tractable problem and belongs to the class *FPT*, and the early name of the theory of parameterized complexity was fixed-parameter tractability.

Many problems have the following form: given an object x and a nonnegative integer k, does x have some property that depends on k? For instance, for the vertex cover problem, the parameter can be the number of vertices in the cover. In many applications, for example when modelling error correction, one can assume the parameter to be "small" compared to the total input size. Then it is interesting to see whether we can find an algorithm which is exponential *only* in k, and not in the input size.

In this way, parameterized complexity can be seen as *two-dimensional* complexity theory. This concept is formalized as follows:

A *parameterized problem* is a language $L \subseteq \Sigma^* \times \mathbb{N}$ where Σ is a finite alphabet. The second component is called the *parameter* of the problem.

A parameterized problem L is *fixed-parameter tractable* if the question "$(x,k) \in L$?" can be decided in running time $f(k) \cdot |x|^{O(1)}$, where f is an arbitrary function depending only on k. The corresponding complexity class is called FPT.

For example, there is an algorithm which solves the vertex cover problem in $O(kn + 1.274^k)$ time , where n is the number of vertices and k is the size of the vertex cover. This means that vertex cover is fixed-parameter tractable with the size of the solution as the parameter.

Complexity Classes

FPT

FPT contains the *fixed parameter tractable* problems, which are those that can be solved in time $f(k) \cdot |x|^{O(1)}$ for some computable function f. Typically, this function is thought of as single exponential, such as $2^{O(k)}$ but the definition admits functions that grow even faster. This is essential for a large part of the early history of this class. The crucial part of the definition is to exclude functions of the form $f(n,k)$, such as n^k. The class FPL is the class with problems solvable in time $f(k) \cdot |x|$ for some computable function f [Grohe, 1999]. FPL is thus a subclass of FPT.

An example is the satisfiability problem, parameterised by the number of variables. A given formula of size m with k variables can be checked by brute force in time $O(2^k m)$. A vertex cover of size k in a graph of order n can be found in time $O(2^k n)$, so this problem is also in FPT.

An example of a problem that is not supposed to be in FPT is graph coloring parameterised by the number of colors. It is known that 3-coloring is NP-hard, and an algorithm for graph k-colouring in time$f(k)n^{O(1)}$ for k=3 would run in polynomial time in the size of the input. Thus, if graph coloring parameterised by the number of colors were in FPT, then P=NP.

There are a number of alternative definitions of FPT. For example, the running time requirement can be replaced by $f(k) + |x|^{O(1)}$. Also, a parameterised problem is in FPT if it has a so-called kernel.Kernelization is a preprocessing technique that reduces the original instance to its "hard kernel", a possibly much smaller instance that is equivalent to the original instance but has a size that is bounded by a function in the parameter. FPT is closed under a parameterised reduction called *fpt-reduction*, which simultaneously preserves the instance size and the parameter. Obviously, FPT contains all polynomial-time computable problems. Moreover, it contains all optimisation problems in NP that allow a Fully polynomial-time approximation scheme.

W *Hierarchy*

The *W* hierarchy is a collection of computational complexity classes. A parameterised problem is in the class $W[i]$, if every instance (x,k) can be transformed (in fpt-time) to a combinatorial circuit that has height at most i, such that $(x,k) \in L$ if and only if there is a satisfying assignment to the inputs, which assigns *1* to at most k inputs. The height thereby is the largest number of logical units with unbounded fan-in on any path from an input to the output. The number of logical units with bounded fan-in on the paths must be limited by a constant that holds for all instances of the problem.

Note that FPT = $W[0]$ and $W[i] \subseteq W[j]$ for all $i \leq j$. The classes in the *W* hierarchy are also closed under fpt-reduction.

Many natural computational problems occupy the lower levels, *W* and *W*.

W

W-complete problems are

- deciding if a given graph contains a clique of size k
- deciding if a given graph contains an independent set of size k
- deciding if a given nondeterministic single-tape Turing machine accepts within k steps ("short Turing machine acceptance" problem)

W

W-complete problems are

- deciding if a given graph contains a dominating set of size k
- deciding if a given nondeterministic multi-tape Turing machine accepts within k steps ("short multi-tape Turing machine acceptance" problem)

W[P]

W[*P*] is the class of problems that can be decided by a nondeterministic polynomial-time Turing-machine that makes at most $O(f(k) \cdot \log n)$ nondeterministic choices in the computation on(x,k) (a k-restricted Turing-machine).Flum & Grohe (2006)

It is known that FPT is contained in W[P], and the inclusion is believed to be strict. However, resolving this issue would imply a solution to the P versus NP problem.

Other connections to unparameterised computational complexity are that FPT equals *W*[*P*] if and only if circuit satisfiability can be decided in time $\exp(o(n))m^{O(1)}$, or if and only if there is a computable, nondecreasing, unbounded function f such that all languages recognised by a nondeterministic polynomial-time Turing machine using f(n)log n nondeterministic choices are in *P*.

Bibliography

Abercrombie, John R. : *Computer Programs for Literary Analysis*, Philadelphia: University of Pennsylvania Press, 2014.

Alon, N. and J. H. Spencer, *The Probabilistic Method*, John Wiley, NJ, 1992.

Ambika, G. : *Computational Aspects In Chaos And Nonlinear Dynamics*, New Age International, Delhi, 2010.

Angeline, P. J.: *Using Selection to Improve Particle Swarm Optimization.* Anchorage, Alaska, USA. 2012.

Atkinson, K.E.: *Elementary Numerical Analysis,* John Wiley, New York, 1993.

Beizer, Boris: *Software Testing Techniques,* International Thomson Computer Press, Boston, MA., 2004.

Bonner, J. T.: *The Evolution of Complexity by Means of Natural Selection.* Princeton: Princeton University Press, 2008.

Bookstein, Fred L.: *Morphometric Tools for Landmark Data.* Cambridge, UK: Cambridge University Press, 2011.

Bracewell, R.N.: *The Fourier Transform and Its Applications,* McGraw-Hill, New York, 1986.

Breiman, L.: *Classification and Regression Trees*, Chapman & Hall, Boca Raton, 1993.

Buckel, W.: *Superconductivity: Fundamentals and Applications,* VCH Publishers, Inc., 1991.

Bulmer, M. G.: *Principles of Statistics*. Mineola, NY: Dover Publications, Inc., 2009.

Chatfield, C.: *Introduction to Time Series and Forecasting*. Springer-Verlag, New York, 1996:

Churchill, R.V.: *Fourier Series and Boundary Value Problems,* McGraw-Hill, New York, 2013.

Clerc, M.: *New Optimization Techniques in Engineering* Springer,Verlag, NY, 2004.

Coulson C. A.: *Supplementary Tables of Molecular Orbital Calculations*, Pergamon Press, Oxford, 2015.

Davis, H.F.: *Fourier Series and Orthogonal Functions*, Allyn and Bacon, Boston, 2013.

Demidenko, E. *Mixed Models: Theory and Applications*. Hoboken, NJ: John Wiley & Sons, Inc., 2004.

Eberhart, R. C., Simpson, P. K., and Dobbins, R. W.: *Computational Intelligence PC tools*, Boston, MA: Academic Press Professional, 1996.

Fisher, R.A.: *Applied Regression Analysis*. Wiley, New York, 2010.

Glazer, A.M.: *Space Groups for Scientists and Engineers*, Academic Press, Boston, 1990.

Hildebrand, F.B.: *Introduction to Numerical Analysis*, McGraw-Hill, New York, 2014.

Hockey, Susan : *A Guide to Computer Applications in the Humanities*, Baltimore: John Hopkins Univerisity Press, 2010.

Kennedy, J. and Eberhart, R. C.: *Particle Swarm Optimization*. Piscataway, NJ. 2015.

Kern, D. Q.: *Process Heat Transfer*, McGraw-Hill, New York, 1950.

Kushilevitz, E. and N. Nisan: *Communication Complexity*, Cambridge, New York, 1997.

Motwani, R. and P. Raghavan, *Randomized Algorithms*, Cambridge, NY, 1995.

Penrose, R.: *Shadows of the Mind*, Oxford University Press, Oxford, 1994.

Pribram, K. H.: *Brain and Perception: Holonomy and Structure in Figural Processing*, Lawrence Erlbaum Assoc, Hillsdale, 2011.

Reichenbach, H.: *The Direction of Time*, University of California Press, Berkeley, 2006.

Scott, P. J.: *Some Aspects of Categories in Computer Science*, Amsterdam, North Holland, 2000.

Shi, Y.: *Swarm Intelligence*, San Francisco: Morgan Kaufmann Publishers, 2001.

Spanier, E. H.: *Algebraic Topology*, McGraw Hill, New York, 1966.

Stapp, H.: *Mind, Matter and Quantum Mechanics*, Springer Verlag, Berlin-Heidelberg-New York, 2013.

Storch, H., and F.W. Zwiers: *Exploratory Data Analysis*. Addison-Wesley, Reading, 1999.

von Neumann, J.: *Mathematische Grundlagen der Quantenmechanik*, Springer, Berlin, 2012.

Weibel-Mihalas, B.: *Foundations of Radiation Hydrodynamics*, Oxford University Press, New York, 2014.

Welsch, R. E. : *Regression Diagnostics*. Hoboken, NJ: John Wiley & Sons, Inc., 2010.

Wood, R.J.: *Categorical Foundations*, Cambridge University Press, Cambridge, 2004.

Index

A

Achievement, 175.
Aggregate Objective Function, 53, 54, 56, 57.
Algal Blooms, 79.
Algorithmic Complexity, 28, 294.
Algorithms, 2, 3, 8, 10, 12, 15, 17, 22, 23, 29, 30, 31, 48, 55, 57, 63, 68, 71, 82, 90, 92, 95, 96, 97, 98, 100, 101, 102, 103, 104, 114, 115, 116, 130, 131, 157, 158, 159, 160, 161, 162, 163, 164, 165, 167, 176, 177, 179, 180, 181, 182, 190, 200, 206, 207, 240, 241, 242, 244, 245, 247, 249, 250, 251, 252, 254, 255, 262, 270, 282, 286, 287, 288, 295, 296, 299, 300, 301, 306, 307, 312, 323, 325, 326, 327.
Altruism Algorithm, 29.
Ant Robotics, 103, 104.
Antenna Design, 115.
Arithmetic Operators, 257.
Arithmetical Hierarchy, 316, 318, 319, 320, 321, 322.
Artificial Immune System, 16, 102.
Artificial Intelligence, 6, 29, 67, 110, 161, 164, 168, 170, 175, 176.

B

Bees Algorithm, 30, 105, 106, 107.
Biological Swarming, 70.
Boolean Circuit, 265, 266.

C

Charged System Search, 31.
Collective Intelligence, 105, 123, 166, 167, 168, 169, 170, 171, 172, 173, 174, 175, 176.
Common Extensions, 95.
Common Misconceptions, 288.
Communication, 26, 32, 33, 34, 68, 75, 83, 86, 92, 94, 104, 106, 122, 123, 169, 171, 181, 239.
Community, 84, 168, 169, 171, 172, 173, 175, 288.
Complexity Classes, 187, 189, 207, 210, 263, 264, 265, 266, 267, 271, 273, 274, 278, 279, 280, 281, 288, 289, 290, 291, 294, 301, 305, 306, 309, 313, 329.
Computational Complexity Theory, 181, 183, 184, 185, 188, 189, 206, 272, 273, 277, 280, 285, 286, 287, 288, 296, 303, 314, 315, 318, 328.
Continuous Complexity Theory, 307.

Contributions, 16, 87.
Coordinate Systems, 42.
Counter Machine Model, 218, 221.
Crowd Simulation, 110, 111.
Cuckoo Search, 18, 31, 162, 176, 179, 180.

D

Decision Problem, 182, 184, 185, 186, 187, 188, 189, 190, 202, 239, 276, 283, 287, 289, 297, 299, 307, 311, 313, 316, 317.
Diffusion Search, 34, 102.
Digital Image Compression, 114.

E

Eagle Strategy, 162, 165.
Evolution Strategy, 162, 164, 165.
Evolutionary Algorithms, 55, 161, 162, 163, 164.
Evolutionary Computation, 17, 25, 28, 161, 162, 163, 164.
Evolutionary Computation Practitioners, 163.
Evolutionary Computing, 131, 162.
Evolutionary Programming, 16, 84, 161, 162, 163, 164.

F

Feasibility Problem, 8.
Firefly Algorithm, 18, 31, 112, 114, 115, 116, 162.
Formal Representation, 62.
Function Problem, 185, 187, 188, 189, 190.

G

Global Convergence, 12.
Glowworm Swarm Optimization, 116.
Governments, 14.
Gradient Descent, 11, 19, 43, 44, 45, 46, 48, 49, 96, 103.

H

Heuristics, 10, 12, 17, 50, 94, 149.

I

Immune Systems, 30, 102, 158, 162.
Insect Migration, 73.

L

Linear Approximation, 39.
Linear Programming Applications, 57.
Linear System, 45, 46, 117.

M

Management, 1, 80, 86, 89, 168, 250.
Mathematical Optimisation, 1, 4, 8.
Mathematical Techniques, 170.
Measurement, 28, 258.
Metaheuristic, 14, 15, 16, 18, 19, 23, 24, 92, 112, 114, 162, 176, 177, 286.
Microeconomic Theory, 60.

O

Operations Research, 14, 57, 106, 119, 330.
Optima, 4, 8, 9, 95, 113, 131, 132, 165.
Optimal Control Applications, 57.
Optimal Input Arguments, 3.
Optimisation Algorithms, 10.
Optimisation Problems, 2, 3, 7, 11, 12, 13, 16, 18, 20, 22, 23, 329.

Optimization Algorithms, 30, 95, 97, 98, 102, 115, 116, 162.

P

Parameter Selection, 21.
Parameterized Complexity, 326, 327.
Pareto Efficiency, 58, 59.
Pareto Frontier, 54, 63, 64.
Pareto Improvements, 58, 60.
Particle Filter, 153, 154.
Particle Swarm Optimization, 25, 31, 32, 52, 55, 69, 70, 100, 102, 114, 115, 132, 133, 162, 177.
Performance, 18, 21, 22, 23, 32, 51, 52, 68, 105, 115, 122, 136, 156, 167, 180, 239, 240, 241, 242, 262.
Policy, 56, 59, 169.
Premature Convergence, 22, 23, 130.
Primitive Recursive Functions, 220, 224, 231.
Probabilistic Polynomial, 322.
Probability Theory, 130, 135, 137, 145, 149.
Programming Applications, 57.
Property, 61, 116, 126, 168, 169, 170, 173, 174, 184, 188, 238, 245, 250, 280, 301, 321, 327.
Protection, 83, 84.

Q

Quantum Computers, 238, 239.

R

Random Access Machine, 206, 212, 216, 225.
Real Variable, 18, 19.
Reductions, 187, 275, 276, 277, 278, 279, 280, 284, 285, 287, 288, 289, 311.
Register Machine Models, 211.
Research, 14, 18, 21, 23, 25, 57, 67, 74, 82, 83, 84, 86, 87, 88, 89, 92, 96, 106, 111, 119, 140, 141, 146, 172, 174, 175, 177, 178, 189, 253, 300, 301, 307, 311.
River Formation Dynamics, 33.

S

Scheduling Problem, 98, 109, 180.
Sequential Importance Resampling, 155, 156, 157.
Sequential Importance Sampling, 157.
Single Aggregate Objective Function, 53.
Social Behavior, 26, 111.
Social Psychologist, 25.
Space Hierarchy Theorem, 271, 272, 273, 274.
Species Swimming, 76.
Stochastic Diffusion Search, 34, 102.
Strong Pareto Optimum, 60.
Structural Design, 115.
Sufficient Conditions, 9.
Swarm Behaviour, 22, 30, 65, 66, 72, 80, 81, 116.
Swarm Intelligence, 19, 29, 34, 35, 67, 68, 70, 81, 82, 92, 100, 102, 116, 162.
Swarm Optimisation, 12, 15, 17, 18, 19, 23.
System Search, 31.

T

Technology, 17, 29, 30, 34, 57, 80, 82, 89, 104, 111, 123,

158, 161, 167, 169, 171, 175, 191, 207.

Time Hierarchy Theorem, 267, 272, 274, 293, 315.

Turing Machine, 104, 150, 189, 190, 191, 192, 193, 194, 195, 197, 198, 199, 200, 201, 205, 206, 207, 208, 209, 210, 211, 214, 215, 216, 218, 223, 229, 230, 235, 236, 237, 239, 263, 264, 267, 268, 269, 270, 271, 278, 279, 282, 292, 295, 302, 303, 304, 309, 311, 313, 314, 321, 322, 324, 325, 329, 330.

Turing Machine Equivalents, 207.

Turing Machines, 191, 192, 196, 197, 199, 200, 201, 206, 207, 208, 210, 211, 214, 215, 237, 263, 264, 265, 267, 292, 303, 304, 307, 315, 321, 324.

V

Vehicle Routing Problem, 99.

W

Welfare, 60.

Wisdom, 123, 175, 176.

□□□